MCQ's ON AGRONOMY

About the Authors

Dr. Kajal Sengupta (born in 1958), Professor of Agronomy, completed his school level education in 1975 and for his brilliant result in Higher Secondary examination he obtained Merit Prize and Merit Certificate. He got National Science Talent Search (N.S.T.S.) Scholarship (under N.S.T.S. scheme 1975) and enjoyed the scholarship throughout his academic carrier - up to Ph.D. level. He also got Merit Certificate for securing 1st class 1st position in M.Sc. (Ag.) Examination. He is a Fellow of Indian Society of Pulses Research and Development, Kanpur, (at I.I.P.R., U.P.). He got Gold Medal Award of Crop and Weed Science Society (CWSS, Nadia, West Bengal).He is former Dean, Faculty of Agriculture and Head, Department of Agronomy, BCKV, West Bengal. He had guided a big number of M.Sc. and Ph.D. students and is presently supervising four M.Sc. and two Ph.D. students. He also acted as the Principal Investigator (PI) of several research projects. He has already published many research papers in various national and international journals and popular extension articles in local papers and magazines / journals. He published more than 25 books and had contributed many book chapters. He regularly participates in various All India Radio (AIR) and *Doordarsan* (TV) programmes. He has a Professional Experience of more than 35 years. He is a Member of Academic Council / Expert / External Examiner of different Universities / Institutes and also a Peer Reviewer of different renowned journals.

Mr. Sukamal Sarkar (born in 1992) is pursuing his Ph.D. in Agronomy at *Bidhan Chandra Krishi Viswavidyalaya* (BCKV), Nadia, West Bengal and CSIRO-Agriculture and Food, Govt. of Australia. He completed his Masters Programme in Agronomy with first class second position in 2016 from BCKV after graduating himself in Agriculture with first class from *Visva-Bharati* in the year 2014. At present he is working as a Senior Research Fellow in the project (CSI4CZ) funded by ACIAR, Govt. of Australia. Recently he is working on development of APSIM-SWIM cropping system model to assess the impact of seasonal salinity and moisture dynamics in Coastal Saline Ecology. Mr. Sarkar won the Alltech Young Scientist Award (2013 and 2015) by Alltech International, USA and "International Plant Nutrition Institute (IPNI) Scholar Awards – 2018". He has the authorship of more than forty number of quality research papers and scientific articles in different journals and magazines of national and international repute. Mr. Sarkar has keen research interest in nutrient management, salinity ecosystem, APSIM-modelling and nutrient dynamics of Agronomic Crops.

Dr. Mahua Banerjee (born in 1976), completed her Higher Secondary education from Durgapur (W.B.), received her B.Sc. (Ag.) Hons. Degree from *Bidhan Chandra Krishi Viswavidyalaya* and completed her M.Sc. (Ag.) and Ph.D. in Agronomy from Indian Agricultural Research Institute (IARI), New Delhi. For her Higher Secondary result she obtained National Scholarship. She stood 3rd in ICAR-JRF examination and got admitted in IARI, New Delhi. She stood 1st in IARI-Ph.D. entrance and on completion of Ph.D. received IARI Merit Medal and G.A. Dastane Award of IARI, New Delhi for outstanding performance in Ph.D. She qualified ICAR-NET (Indian Council of Agricultural Research - National Eligibility Test) in Agronomy in 2001 and 2005. She worked as Senior Research Fellow in a Central Pollution Control Board sponsored research Project in the Division of Environmental Science of IARI. She stood 12th in the West Bengal Civil Service Exam. of 2004 and worked in the capacity of Deputy Magistrate and Deputy Collector from 2005 to 2007. She was also selected in West Bengal Agricultural Service (Administrative). She joined *Visva-Bharati* as Lecturer in Agronomy in 2007. She had already guided a considerable number of M.Sc. and Ph.D. students and is presently supervising three M.Sc. and four Ph.D. students. She also acted as the PI of a CSIR project, IPNI sponsored project and is acting as PI of several research projects. She has already published many research papers and popular articles in various national and international journals and had already contributed about 10 chapters for publication as book chapter. She regularly participates as resource person in various radio and TV programmes.

Miss Purabi Banerjee (born in 1994), a young promising agronomist, is currently pursuing her Ph. D. degree at *Bidhan Chandra Krishi Viswavidyalaya*, Nadia, West Bengal. She got her B.Sc.(Ag.) Hons. Degree in 2017 and M.Sc. (Ag.) in Agronomy in 2019 from *Bidhan Chandra Krishi Viswavidyalaya*, West Bengal. She poses an excellent academic record. She received State Government fellowship for Ph. D. research. She has already published a few research papers in referred (peer reviewed) journals and popular extension articles. Agronomy of pulse crops, plant nutrition and agro-meteorology are her main area of research.

MCQ's ON AGRONOMY

For paramount importance to students appearing for interviews, viva-voce, comprehensive and other competitive examinations like

IAS, IFS, ARS, NET, PCS, CSC, IARI, SAUs

Entrance Test and Banking Services of Agricultural Subjects

Kajal Sengupta
Sri Sukamal Sarkar
Mahua Banerjee
Purabi Banerjee

New Delhi – 110 034

NEW INDIA PUBLISHING AGENCY

101, Vikas Surya Plaza, CU Block, LSC Market
Pitam Pura, New Delhi – 110 034, India
Email: info@nipabooks.com
Web: www.nipabooks.com

For customer assistance, please contact

Phone: + 91-11-27 34 17 17 Fax: + 91-11- 27 34 16 16
E-Mail: feedbacks@nipabooks.com

ISBN 978-93-89571-95-0

Composed and Designed by NIPA.

Bidhan Chandra Krishi Viswavidyalaya
P.O. Krishi Viswavidyalaya, Mohanpur-741252
District: Nadia, West Bengal, India
Website: www.bckv.edu.in

No. VC/BCKV/114/270
Date: 18.12.2019

Tel. : 033-25879772
03473-222666
Fax: 03473-222275
Email: bckvvc@gmail.com
ddpatra@rediffmail.com
Cell : + 91-9830071278

Dr. D.D. Patra, Ph D (IARI, New Delhi), FNA Sc., FISSS, FNAAS, FWAST
Vice-Chancellor

Foreword

In Agriculture, Agronomy is the mother subject which deals essentially with all basic subjects of agriculture. Agronomy is the study of how plants and soil can best be used for food, fuel, fiber, fodder and soil reclamation. The importance of agronomy combines the studies of ecology, biology, chemistry, soil science, and genetics to determine the capabilities of various types of plants and soils in their environments.

The book '**MCQ's on Agronomy**" written by **Prof. Kajal Sengupta, Sri Sukamal Sarkar, Dr. Mahua Banerjee and Ms. Purabi Banerjee** is a good one, as it deals with Agronomy in totality with a focus on the syllabi of the different agricultural universities as well as competitive examinations like university entrance tests, NET, ARS, IFS, bank examinations, etc. The book will serve as an excellent ready reckoner for students, researchers and teachers.

This book is very precisely written by the authors and covered most of the aspects in Agronomy and will fulfill the requirements of aspirants of agricultural jobs.

I congratulate and complement the authors for their sincere efforts in bringing out the book. I hope, the book will gain popularity among the students, teachers and other readers associated with agriculture.

I wish a great success of this publication.

(D.D. Patra)

Preface

Agronomy, one of the most important branches of Agriculturc, assumed newer dimensions to provide favourable environment to the crop for higher productivity. The importance of agronomy combines the studies of different allied subjects like ecology, biology, chemistry, soil science, genetics, etc., to determine the capabilities and potentialities of various types of plants in their aerial and edaphic environments.

Now-a-days, evaluation of the students in different type of competitive examinations is done by setting multiple choice questions (MCQs) and at present it is one of the prime methods for evaluation. The evaluators prefer MCQs for quick and thorough evaluation. From this context, we took the onus to make such objective book in which plenty of recent information in the form of solved query are gathered that makes the students more informative and enthusiastic in achieving their goal towards the newest topic. The experience shows that only a few books are available to most of the students; on the other hand if available they remain unread, either due to lack of time or inadequate information as per their requirement or demand. Therefore, to cater the needs of students competing for various examinations / interviews the book "MCQs ON AGRONOMY" has been written. The book contains the most important and useful information about Agronomy.

The book, **MCQ's on Agronomy,** contains more than two thousand (2000) MCQ, covering all fields of Agronomy such as agro-meteorology, crop management, soil management, nutrient management, water management, weed management, dryland agriculture, agro-forestry, etc. We have tried to cover all branches of Agronomy. The objective questions with multiple choices have been framed and arranged in such a way so that the students can prepare themselves for different type of examinations. The questions are framed purposefully in a manner so that the entire main topics are covered step by step. No doubt that this book would be of paramount importance for students appearing for interviews, viva-voce, comprehensive and other competitive examinations like IAS, IFS, ARS, NET, PCS, CSC, IARI, SAUs Entrance Test and Banking Services of agricultural subjects, especially in Agronomy as it would help them to revise the entire agronomy within a short time and recollect the vast knowledge they gained in their academic career and also by reading different books of Agronomy. We hope that the book will gain popularity among the students, teachers and other readers in agricultural lines and serve as a comprehensive reading material for them.

Authors

Contents

Questions

Agro-meteorology, Crop Management, Soil Management, Nutrient Management, Water Management, Weed Management, Dryland Agriculture and Agroforestry

1. Permissible biuret level in most urea fertilizer is

   a) 5% b) 4%

   c) 3% d) 2%

2. The trade name of fernoxone is -

   a) 2,4 – D amine b) 2,4 – D ethyl ester

   c) 2,4 – D – Na salt d) 2,4,5 – T

3. Boric acid contains_________ % Boron

   a) 13 b) 15

   c) 17 d) 19

4. Which one is the tailor made fertilizer for pulse crops?

   a) Urea b) Ammonium nitrate

   c) DAP d) CAN

5. Basic slag contains ____________ % P_2O_5

   a) 10 b) 14

   c) 20 d) 25

6. Yuan Longping is consider as the Father of

   a) Tillage b) Hybrid rice

   c) Organic rice d) Organic farming

7. Water use efficiency is highest in which crop?

   a) Wheat b) Potato

   c) *Ragi* d) Rice

8. Which one is not a C_4 plant?

   a) Maize b) Sugarcane

   c) Agave d) Sorghum

9. *Acetobacter* inoculation for N-fixation is used for
   a) Wheat b) Potato
   c) Sunflower d) Sugarcane
10. Zero tillage system was first successfully used in
   a) Japan b) China
   c) USA d) Australia
11. The concept of LER is given by
   a) Menegay b) Mead & Willey
   c) Jackson d) Chaung
12. The line connecting points of equal solar radiation is known as
   a) Isohume b) Isohel
   c) Isoneph d) Isohyet
13. The Brown box dots on food package indicates
   a) Egg product b) Meat pro1duct
   c) Organic product d) Pure vegetarian product
14. The organization related to export of Agricultural processed products is
   a) NABARD b) ICAR
   c) APARI d) APEDA
15. '**Annals of Agricultural Research**' is published from
   a) ICAR b) IARI
   c) IIPR d) NBRI
16. **Phyllody** is a common problem in
   a) Rice b) Niger
   c) Castor d) Sesame
17. ***Arhar*** (Pigeon pea) is categorised as
   a) Cross pollinated b) Self pollinated
   c) Often cross pollinated d) Often self pollinated
18. The International Institute of Tropical Agriculture is located at
   a) India b) Indonesia
   c) Japan d) Nigeria

19. The crop that has the largest area under transgenic globally is

a) Rice b) Soybean

c) Wheat d) Cotton

20. Sunflower is known as an indicator plant for the deficiency of

a) N b) K

c) B d) Zn

21. The critical period of crop-weed competition is approx. first ______ days for most of the field crops

a) 15 b) 20

c) 25 d) 40

22. Zero tillage has been found successful in ___________ crop

a) Wheat b) Sugarcane

c) Maize d) Potato

23. The world most problematic weed is

a) *Cynodon dactylon* b) *Cyperus rotundus*

c) *Eleusine indica* d) *Tridex procumbens*

24. The National Research Centre for groundnut is located in

a) Jamnagar b) Jaipur

c) Jodhpur d) Junagadh

25. For good germination of wheat seed the optimal temperatures (°C) range from

a) 5-10 b) 15-25

c) 30-35 d) 35-40

26. Optimum period of sowing of ***capsularis*** jute is

a) January-February b) March-April

c) May-June d) July-August

27. ***Ufra*** disease of rice is

a) Due to attack on nematodes b) Due to deficiency of N

c) Due to deficiency of Zn d) Due to attack of bacteria

28. Canola seed contains less than ________ of erucic acid

a) 15 b) 10

c) 5 d) 2

29. Concept of plant ideotype was given by
 a) Meyer b) Swaminathan
 c) Donald d) Smith

30. Fertilizer prone to fire hazard is
 a) Urea b) CAN
 c) DAP d) Ammonium nitrate

31. **White bud** of Maize is caused by _________ deficiency
 a) Zn b) B
 c) Mn d) Mo

32. The first KVK was established in the year _________ at Pondicherry/ (Puducherry)
 a) 1975 b) 1974
 c) 1973 d) 1972

33. ____________________________ is the purest form of seed
 a) Foundation seed b) Registered seed
 c) Certified seed d) Nucleus seed

34. The best temperature for tuberization in potato is
 a) 10°C b) 15°C
 c) 20°C d) 30°C

35. Which one is a contact non-selective herbicide?
 a) Propanil b) Paraquat
 c) Glyphosate d) Atrazine

36. Example of beneficial mineral element for plants
 a) Si b) Mn
 c) Cu d) Fe

37. The recommended depth of sowing of wheat is
 a) 2 cm b) 5 cm
 c) 8 cm d) 10 cm

38. Rock phosphate is recommended for
 a) Alkaline soil b) Acidic soil
 c) Neutral soil d) Saline soil

39. Pulse crop which generally does not fix atmospheric N is
    a) Pigeon pea b) Lentil
    c) Grass pea d) Rajmash

40. 'Indian Society of Agronomy' was established in the year
    a) 1950 b) 1955
    c) 1960 d) 1965

41. Which one of the following is not related with sugarcane?
    a) SBI b) IISR
    c) CSIR d) NSI

42. Which one of the following crop is not included in research programme of **ICRISAT**?
    a) Sorghum b) Pigeon pea
    c) Chick pea d) Chickling pea

43. Fertilizer which supplies three (3) essential plant nutrients is
    a) SSP b) MOP
    c) DAP d) SOP

44. ***Akiochi*** disease in rice occurs due to the toxicity caused by ________
    a) Zn b) Fe
    c) P d) H_2S

45. Multi-storied cropping system is commonly practiced in
    a) Kerala b) Bihar
    c) Punjab d) Madhya Pradesh

46. Percentage of edible oil in rice bran is
    a) 5-8% b) 10-15%
    c) 18-20% d) 22-25%

47. Wheat protein is known as
    a) Dhurin b) Zein
    c) Gluten d) Ricin

48. The seed rate of true potato seed is__________ g ha^{-1}
    a) 50-80 b) 100-150
    c) 180-220 d) 250-300

49. Chemical recommended to break the dormancy of potato is -
   a) Thiourea b) IAA
   c) IBA d) NAA

50. The ratio of cotton seed to lint is -
   a) 1 : 2 b) 1 : 3
   c) 2 : 1 d) 3 : 1

51. Concept of pH was given by
   a) Silen b) Schofield
   c) Sorensen d) Israelson

52. The contribution of agriculture in GDP is
   a) 23% b) 30%
   c) 40% d) 18.5%

53. Which of the following transgenic crop has maximum cultivated area in the world?
   a) Maize b) Soybean
   c) Rice d) Cotton

54. Flvr savr is a variety of
   a) Tomato b) Potato
   c) Brinjal d) Rice

55. The statistical test is used to determine the goodness of fit is
   a) Chi-square test b) Z test
   c) F test d) T test

56. The biogas has major content with CO_2 is
   a) C_2H_2 b) CO
   c) CH_4 d) N_2

57. *Utera* cultivation is mainly practices in
   a) Uttaranchal b) Chhattisgarh
   c) MP d) Bihar

58. The most common method of hybridization for wheat is.
   a) Floral dip method b) Open stigma method
   c) Hot water dips d) Gogo method

59. In cereal grains, starch consists
   a) Amylose b) Amylopectin
   c) Both a & b d) Cellulose

60 Hatch and slack pathway found in
   a) Wheat b) Maize
   c) Rice d) Soybean

61. The net cultivated area in India is
   a) 143 mha b) 150 mha
   c) 320 mha d) 180 mha

62. The Wheat and Maize Improvement Centre is located at
   a) Mexico b) Philippines
   c) Karnal d) USA

63. The soil amendment mainly used for reclamation of sodic soils is
   a) Pyrite b) Lime
   c) Gypsum d) Pressmud

64. The citrus canker disease of lemon was introduced in India from
   a) USA b) Mexico
   c) Argentina d) Japan

65. In pulse crop which fertilizer is advice to more basal application
   a) P b) N
   c) K d) Ca

66. Optimum temperature for cool season crops
   a) 10-15°C b) 15-20°C
   c) 20-25°C d) 25-30°C

67. The region in India , where Boron deficiency is mainly occurs
   a) North-west region b) Central region
   c) Coastal region d) Southern region

68. The Agriculture and allied commodity has maximum contribution to GDP
   a) Milk b) Rice
   c) Wheat d) Sorghum

69. Cotton leaf curl virus is transmitted by

a) Aphid
b) Whitefly
c) Jassid
d) Leaf hopper

70. Loose smut disease of wheat is a

a) Seed borne
b) Soil borne
c) Air borne
d) Water borne

71. Which of following is not a green house gas?

a) CO_2
b) O_3
c) N_2O
d) N_2

72. The first Director General of ICAR

a) Dr. B.P. Pal
b) Dr. B. Vishwanath
c) Dr. M.S.Swaminathan
d) Dr. Mangla Rai

73. The political personality who got Norman Borlaug award

a) C. Subramanium
b) Lal Bahadur Shastri
c) J.L. Nehru
d) Sharad pawar

74. The Brown box dots on food package indicates

a) Organically produce product
b) Meat and animal produce product
c) Vegetarian product
d) Egg product

75. Which state has maximum area in the fruit production?

a) UP
b) Gujarat
c) Maharastra
d) MP

76. The organization is related to export of Agricultural processed products

a) APEDA
b) APARI
c) NABARD
d) CACP

77. IVLP stands for

a) Integrated Village Linkage programme
b) Institute Village Linkage programme
c) Integrated Village Linkage project
d) Institute Village Linkage project

78. Land resources in India as a whole world is

a) 2%
b) 10%
c) 12%
d) 16%

79. The cartigen quato protocol for Agriculture product was come into existence in the year of

a) 1998 b) 2003

c) 2004 d) 2005

80. The soil order *vertisol* is related to

a) Black soil b) Laterite soil

c) Red soil d) Alluvial soil

81. Which vegetable has maximum source of vitamin A

a) Radish b) Carrot

c) Turnip d) Spinach

82. Mango hybrid 'Ratna' is a cross of

a) Deshari x Neelam b) Neelam x Deshari

c) Neelam x Alphanso d) Deshari x Alphanso

83. The propagation method 'Layering' is mainly done in

a) Mango b) Lemon

c) Sweet orange d) Banana

84. The established Tractor company in India was

a) Hindustan b) HMT

c) Eicher d) Ford

85. The plants which have maximum water use efficiency are

a) C_4 b) C_3

c) CAM d) None

86. The largest living animal on the earth is

a) Elephant b) Blue whale

c) Lion d) Buffalo

87. The International Institute of Tropical Agriculture is located at

a) Syria b) Nigeria

c) Japan d) India

88. Which of the following is a goat breed

a) Suri b) Surthi

c) Lohi d) Nagori

89. The gestation period of sheep is

a) 165 days b) 147 days

c) 250 days d) 300 days

90. The world's best quality wool is produced from sheep breed

a) Marino b) Gaddi

c) Chokla d) Bikaneri

91. Discharge formula from 90° V-notch type of weir when head (H) in cm is

a) $Q = 138\ H^{2.5}$ b) $Q = 0.0138\ H^{2.5}$

c) $Q = 0.138\ H^{1.5}$ d) $Q = 0.0138\ H^{3.5}$

92. Which type of plough makes V shaped furrow is

a) Mould board plough b) Desi plough

c) Ridge plough d) Disc plough

93. The common multi purpose tree spp. (MPTs) is

a) Albizia lebbeck *b) Cassica siamea*

c) Gliricidia sepium d) Eucalyptus

94. The fast growing tree spp. is

a) Eucalyptus b) Banyan

c) Neem d) Cassia

95. The Agri-Silvi-Pastoral system has

a) Crop+Animal b) Crop+Tree

c) Crop+Tree+Pasture+Animal d) None

96. The staple length of fine quality cotton fibres in micronaurie is

a) 3.0-3.9 b) 4.0-4.9

c) 5.0-5.9 d) None

97. The prescribed rules which traditionally approved by society

a) Custom b) Beliefs

c) Rituals d) Folk ways

98. The consciousness & purposeful action by the peoples is known as

a) Motivation b) Co-ordination

c) Skill d) Education

99. The Indian Economy is a type of

a) Socialist | b) Capitalist
c) Mixed | d) None

100. The short term loan which given to farmers by the Banks is also known as

a) Soft loan | b) Crop loan
c) Term loan | d) Hard loan

101. Method showing the worth & value of New practices is a

a) Result demonstration | b) Method demonstration
c) Observation plot | d) Field demonstration

102. The Institution which does provides loan to the farmers is

a) Commercial Banks
b) Primary cooperative societies
c) RRB's
d) Primary land development societies

103. The Intellectual Development age of children is

a) Birth to 4 years | b) 6-10 years
c) 10-12 years | d) 14-16 years

104. The calories requirement for children of age group 6-11month

a) 120 cal. | b) 115 cal.
c) 110 cal. | d) 105 cal.

105. The practical limit for suction head of centrifugal pump

a) 10m | b) 7m
c) 5m | d) 2m

106. The K-fixation is maximum in

a) Vermiculite | b) Illite
c) Chlorite | d) Mica

107. The fertilizer nutrient which immediately fixed after application in the soil

a) N | b) P
c) K | d) Ca

108. The maximum production of fish in the state of

a) Maharastra b) Gujarat

c) West Bengal d) UP

109. The maximum per capita consumption of fish in the world is in

a) Japan b) USA

c) Russia d) India

110 The sex linked gene are found on the

a) X-chromosome b) Y-chromosome

c) S-chromosome d) Z-chromosome

111. The chromosomes which are either differs in numbers or morphology in the male & female are called

a) Sex chromosomes b) Autosomes

c) Euploidy d) Accessory chromosomes

112. The mitigating stress with high degree of tolerance is called

a) Drought escaping b) Drought tolerance

c) Drought resistance d) Drought avoiding

113. The sugars present in DNA & RNA are

a) Glucose & Mannose b) Lactose & sucrose

c) Deoxyribose & Ribose d) Glucose & Maltose

114. The longest phase in mitosis is

a) Prophase b) Telophase

c) Anaphase d) Metaphase

115. The unsaturated fatty acid is determined by

a) Saponification No. b) Iodine No.

c) Both d) None

116. The suitable wheat species for restricted irrigation is

a) *Triticum aestivum* b) *Triticum durum*

c) *Triticum dicoccum* d) *Triticum vulgare*

117. The Casparian strip is present in

a) Epidermis b) Endodermis

c) Cortex d) Root hair

118. The green house effect is cause due to
   a) Increase in concentration of CO_2 in atmosphere
   b) Incoming UV rays on the earth
   c) Depletion of ozone layer
   d) Absorption of out going long wave radiation

119. The Acid rain is mainly due to the presence of atmospheric gas
   a) CO_2 b) SO_2
   c) NO_2 d) NO_2 & SO_2

120. Cartilages are
   a) Muscular tissues b) Epithelium tissues
   c) Connective tissues d) None

121. Most suitable crop for the late onset of monsoon to reduced the risk of crop failure is
   a) Tur/Arhar b) Soybean
   c) Sunflower d) Rice

122. Fruits are generally
   a) Acidic b) Alkaline
   c) Neutral d) None

123. The freezing point depression of a chemical solution is measured by
   a) Calorimeter b) Viscometer
   c) Osmometer d) None

124. One cusec water discharged for one hour is equal to..........litres
   a) 101×10^3 b) 106×10^4
   c) 125×10^2 d) 101 x 102

125. The chemical used for preventing the fruit drop in grape fruit is
   a) NAA 20 ppm b) GA_3 50 ppm
   c) NAA 100 ppm d) None

126. The computer software used for soil management strategy and natural occurrence is
   a) Gossym b) Comex
   c) Wofex d) None

127. 'Arrowing' is occurs in
   a) Sugercane b) Rice
   c) Gram d) Wheat

128. The insect pest which cause huge destruction in sugarcane in Maharastra recently
   a) Pyrilla b) Woolly aphid
   c) Stem borer d) Top borer

129. Which of the following is not a fish
   a) Bombay duck b) Silver fish
   c) Milk fish d) Dog fish

130. The average annual human labour power in India is
   a) 0.1KW b) 0.5KW
   c) 100KW d) 10KW

131. The nitrogen use efficiency in irrigated condition can be enhanced by application of fertilizer at
   a) Deep placement b) Split dose
   c) Fertigation d) None

132. The safe storage moisture percentage for wheat is
   a) 2-5 % b) 5-10 %
   c) 10-14% d) 14-16%

133. The atmospheric phenomenon which occur at regular interval and cosmologically controlled is
   a) Tsunami b) Earthquake
   c) Ocean tide d) El-Nino

134. Most effective instrument for measuring soil water content in-situ is
   a) Tensiometer b) Neutron probe
   c) TDR d) Gypsum block

135. Biochemical found in fish which is responsible for reduction in cardiovascular disease
   a) Poly unsaturated fatty acid (PUFA)
   b) Trimethylamine
   c) Total volatile based nitrogen
   d) Omega 3-fatty acid

136. Rice gene bank is situated at
  a) Philippines b) Japan
  c) China d) India
137. The plant cut at 2 meter height is known as
  a) Pollarding b) Brashing
  c) Brushing d) Hedging
138. Rainfed area of India is about
  a) 57 Mha b) 30 Mha
  c) 65 Mha d) 40 Mha
139. The National Research Centre on Groundnut is located at
  a) Junagarh b) Bikaner
  c) Delhi d) Jabalpur
140. Ozone layer in the atmosphere absorbs
  a) Infra red radiation b) Ultra-violet radiation
  c) Total solar radiation d) All
141. The minimum juice content in squash
  a) 15% b) 22%
  c) 25% d) 30%
142. The minimum TSS / brix in preserve should be
  a) 68% b) 60%
  c) 72% d) 80%
143. The viability of onion seeds is of .... duration
  a) 1 year b) 2 year
  c) 3 year d) >3 year
144. Paizi (wild onion) is weed of.......... season
  a) *Rabi* b) *Zaid*
  c) *Kharif* d) None
145. The sowing time in nursery of spring / summer tomato is
  a) Nov-Dec b) Jan-Feb
  c) March-April d) June-July

146. Which one of the Japanese style garden
  a) Taj garden b) Shalimaar garden
  c) Budha jayanti park d) Rastrapati garden
147. The winter season annuals are sown in the nursery in the month of
  a) Jan- Feb b) Sept-Oct
  c) March-April d) June-July
148. Calculate plant population of lime in 1 ha if spacing is 7 x 7 metre
  a) 204 b) 304
  c) 300 d) 200
149. The most intensive system of vegetable cultivation is
  a) Market gardening b) Truck gardening
  c) Kitchen gardening d) Garden seed production
150. The term monsoon is derived from an Arabic word
  a) Mausim b) Mosam
  c) Mossom d) None
151. *Khaira* disease of rice is due to.............. deficiency
  a) Zn b) Mo
  c) Cu d) Mn
152. The equivalent acidity of ammonium sulphate is
  a) 110 b) 128
  c) 148 d) 80
153. The biuret content in urea for foliar fertilization should be
  a) 0.5% b) 1%
  c) 1.5% d) 2%
154. The suitable fertilizer for acidic soils
  a) $CaCN_2$ b) Ammonium sulphate
  c) Ammonium phosphate d) Ammonium chloride
155. Zinc phosphide is used for the control of
  a) Rats b) Birds
  c) Snakes d) Fishes

156. The first animal on earth is

a) Arthropoda b) Protozoa

c) Insecta d) Echinochordata

157. The endosperm of Angiosperm is

a) Haploid b) Diploid

c) Triploid d) Polyploid

158. The starter solution is of

a) N b) P

c) K d) NPK

159. The particle size in saltation is

a) <0.1 mm b) 0.1-0.5 mm

c) 0.5-1.0 mm d) 1.0-2.0 mm

160 Which on e of the following is taken into consideration for IPM

a) Cropping system b) Agro-eco system

c) Agroforestry d) None

161. The inter cropping system are profitable when LER is

a) 0.5 b) 0.75

c) 1.0 d) 1.25

162. Which is most profitable for dry land agriculture is

a) Mixed cropping b) Inter cropping

c) Crop rotation d) Relay cropping

163. The Molya disease of barley is caused by

a) *Heterodera avanae* b) *Longidera* sp.

c) *Meliodogyne* spp. d) *Pratylenchus* spp.

164. The barley rust is caused by

a) Puccinia hordei b) Puccinia recondite

c) Tilitia foeatida d) Puccinia gramminis

165. Tundu disease of wheat is caused by

a) Nematode b) Bacteria

c) Both a & b d) Fungus

166. Loose smut of wheat is
a) Internally seed borne
b) Externally seed borne
c) Air borne
d) Soil borne

167. The component of Blitox-50 and Blifax are
a) Copper sulphate
b) Copper oxychloride
c) Zinc sulphate
d) None

168. Fumigation is done to control
a) Pink boll worm
b) Jassids
c) Aphids
d) None

169. Hot water treatment is given to
a) Soil
b) Planting material before sowing
c) Both a & b
d) None

170. The optimum time of budding in aonla is
a) June-July
b) Feb-March
c) July-Aug
d) None

171. The spray used for the control of pre mature leaves, flower and fruit drop is
a) IAA @200 ppm
b) NAA @ 200 ppm
c) IBA @ 200 ppm
d) None

172. The soil water erosion is mainly caused by rain drops is
a) Splash
b) Sheet
c) Rill
d) Gully

173. White bud of Maize is caused by
a) Mn deficiency
b) Zn deficiency
c) Cu deficiency
d) Mo deficiency

174. Iron deficiency symptoms are
a) Yellowing of older leaves
b) Yellowing of young leaves
c) Yellowing of middle leaves
d) None

175. Which type of be hives are commonly used in India?
a) Thompson type
b) Landstein type
c) Newton's type
d) None

176. The main advantage of organic farming is
   a) Subsistence production   b) Sustainable production
   c) Both   d) Max production
177. Indian Lac Research Institute is located at
   a) N. Delhi   b) Ranchi
   c) Karnal   d) Varanasi
178. Central Marine Fisheries Research Institute is located at
   a) Orissa   b) Cochin, Kerala
   c) AP   d) WB
179. Project Directorate on Biological Control is located at
   a) Bangalore   b) Hyderabad
   c) N. Delhi   d) Mumbai
180. Boron deficiency in leaf starts from
   a) Base to tip   b) Tip to base
   c) Either a or b   d) At the margin
181. The mosquito eating fish is
   a) Catla   b) Gambussia
   c) Telapia   d) Both c and d
182. The best time of harvesting fishes is
   a) Summer   b) Spring
   c) Winter   d) All
183. The length of fish fingerlings is
   a) 20 mm   b) 30 mm
   c) 40 mm   d) 10 mm
184. Milk fever is caused by
   a) Mn deficiency   b) Ca deficiency
   c) Fe deficiency   d) None
185. Which one of the following is a bacterial disease of lactating animals
   a) FMD   b) Rinderpest
   c) Mastitis   d) Anaemia

186. The open space required for a cross breed cow

a) 90-100 ft^2
b) 100-110 ft^2
c) 120-130 ft^2
d) 150-170 ft^2

187. The gestation period of a cow is

a) 282 days
b) 310 days
c) 272 days
d) 292 days

188. Which breed of buffalo is continuously calving and lactating?

a) Murrah
b) Bhadawari
c) Surati
d) Neeli

189. The first KVK was established in the year at Pondicherry

a) 1972
b) 1974
c) 1976
d) 1978

190. TRYSEM was started in

a) 1978
b) 1979
c) 1966
d) 1980

191. NABARD was established in the year

a) 12 July, 1982
b) 15 Aug, 1979
c) 2 Oct, 1980
d) None

192. The digestion in poultry occurs in

a) Small intestine
b) Large intestine
c) Vent
d) Gizzard

193. White leg horn is originated from

a) Italy
b) USA
c) Australia
d) New Zealands

194. Which is purest category of seed?

a) Breeder seed
b) Foundation seed
c) Registered seed
d) Nucleus seed

195. In transamination, ... is used as a base material

a) NH_3
b) Glutamic acid
c) Glutamin
d) A ketoglutaric acid

196. The topping in tobacco means

a) Inflorescence with 2-3 leaves b) Upper 2-3 leaves

c) Lower 3-4 leaves d) All

197. The best temperature for tuberization in potato is

a) 15°C b) 18.4°C

c) 25°C d) 30°C

198. The sugarcane maturity is indicated by a brix ratio of

a) 15 b) 20

c) 25 d) 30

199. 52. Which one is a contact non-selective herbicide

a) Paraquat b) Glyphosate

c) Propanil d) Atrazine

200. The number of steps in teaching methods of extentions are

a) 5 b) 6

c) 4 d) 3

201. The % of innovaters in adoption process

a) 2.5% b) 13.5%

c) 16% d) 33%

202. The best method for showing the result of a new practice is

a) Result demonstration b) Method demonstration

c) Both d) None

203. Extension education is a

a) Change in skill b) Change in knowledge

c) Change in attitude d) All

204. The best method for selecting a local leader by an extension worker is

a) Sociometriy b) Election

c) Both d) None

205. The first school of a child is

a) Caste b) Family

c) Religion d) Society

206. The term Chrmosome was given by Waldayer in

a) 1888 b) 1889

c) 1881 d) 1880

207. The word genotype was given by Johanson in

a) 1903 1905

b) 1900 c) 1901

208. The definition of test cross is

a) F_1 x Both parent b) F_1 x dominant parent

c) F_1 x Recessive parent d) None

209. The mutation at gene level is known as

a) Gene mutation b) Chromosome mutation

c) Point mutation d) All

210 When the offspring is superior than mid parent is called

a) Heterobaltiosis b) Hybrid vigour

c) Both d) None

211. The inventory of farm resource is .... done in a year

a) One b) Two

c) Three d) Four

212. The definition of demand in economic term is

a) Willingness to purchase b) Ability to pay

c) Both d) None

213. Anything which has value

a) Goods b) Transferability

c) Scarcity d) Utility

214. If marketing efficiency is increased than the farmers profit is

a) Decreased b) Increased

c) Static d) None

215. The Total physical product (TPP) in IIIrd zone is

a) Increased b) Decreased

c) Increased and decreased d) None

216. The second zone in production function is

a) Rational b) Irrational

c) Both d) None

217. The coding of three base pair is known as

a) Codon b) DNA

c) Amino acids d) None

218. The hormone responsible for let down of milk

a) Parathormone b) Adrenalins

c) Oxitocin d) Thyroxin

219. The good source of animal protein is

a) Bone meal b) Fish meal

c) Skimmed milk powder d) None

220. The value of Karl Pearson's correlation coefficient is

a) ± 1 b) 0 -1

c) -1 to +1 d) - " to +"

221. Which method of central tendency shows mid value?

a) Median b) Mode

c) Arithmetic mean d) Harmonic mean

222. The CV is calculated by

a) SD/Mean X 100 b) Mean/SD

c) SD X Mean d) (SD – Mean) X 100

223. The replication and randomization are used in which design

a) CRD b) RBD

c) LSD d) SPD

224. Example of supplementary enterprises are

a) Poultry industry and farm b) Steel industry and farm

c) Both d) None

225. In cross pollinated crops, selfing increases

a) Homogygosity b) Heterogygosity

c) Homogeneity d) All

226. G J Mendel worked on

a) Garden pea
b) Wild pea
c) Field pea
d) None

227. Cyst forming nematode is

a) Ditylenchus
b) Pratylenchus
c) Longidorus
d) Heterodera

228. Ergot of Bajra produces

a) Scleoratia
b) Oomycelia
c) Claviceps purpurea
d) Triticum

229. Triticale (First man made cereal) is a cross between

a) Rye x Wheat
b) Rye x Wheat
c) Barley x Wheat
d) Barley x Rye

230. If hypothesis is true but rejected , which error is

a) Type I error
b) Type II error
c) Both
d) None

231. Which disease is of not common occurrence in India

a) Panama wilt
b) Cocoa disease
c) Spongy tissue
d) Citrus cancker

232. Sexual attraction of mature insects

a) Hormones
b) Chemosterilents
c) Pheromones
d) Enzymes

233. Which is the seed born nematode

a) Cyst nematode
b) Needle nematode
c) Spiral nematode
d) None

234. % P formula is

a) $\%P_2O_5$ x 0.43
b) $\%P_2O_5$ x 0.043
c) $\%P_2O_5$ x 43
d) $\%P_2O_5$ x 4.3

235. Father of binomial classification is

a) C. Linneus
b) Mendal
c) De Vris
d) Lamark

236. The fibre length of long staple cotton is
   a) 20.5-24.0 mm   b) >24.5 mm
   c) 27.5-32.5 mm   d) > 20.5 mm

237. Baraderi is an important feature of
   a) Mughal garden   b) Japanese garden
   c) English garden   d) Egyptian garden

238. *Gnorimoschema opercullela* is
   a) Potato tuber moth   b) Late blight
   c) Cut worm   d) Potato scab

239. Black heart' disease is physiological disorder of
   a) Tomato   b) Chilli
   c) Cabbage   d) Potato

240. The quickest method of plant breeding is
   a) Selection   b) Hybridization
   c) Mutation breeding   d) Introduction

241. 'Self pollinated' species known as
   a) Allogamous   b) Cleistogamous
   c) Autogamous   d) Chasmogamous

242. The average annual rainfall of 250-500 mm comes under the area
   a) Arid   b) Semi arid
   c) Sub humid   d) Humid

243. Which one of the following potato varieties is of early duration
   a) Kufri Jyoti   b) Kufri Ashoka
   c) Kufri Sinduri   d) Kufri Badsa

244. *Khaira* disease in rice is first noted in the state of
   a) UP   b) West Bengal
   c) Himachl Pradesh   d) Punjab

245. Which one of the Characteristics helps to identify the alkali soil
   a) White /ash colour on soil surface
   b) Redish colour on soil surface
   c) Black spot on soil surface
   d) None

246. The equation log (A-Y)=Log A-Cx is

a) Leibig's equation
b) Spilman's equation
c) Mitscherlich equation
d) Gapories equation

247. The 1st step of nitrogen mineralization process is

a) Ammonification
b) Immobilization
c) Amminization
d) Nitrification

248. The head quarter of CMMYT is situated

a) Philippine
b) Vietnam
c) Indonesia
d) Mexico

249. When soil pH changes from 7.0-6.0, its H-ion conc increases by times

a) 1
b) 10
c) 100
d) 1000

250. If the component crop in the cropping system is equal, the aggressivity index will be

a) 0
b) -1
c) 1
d) 2

251. Triticum aestivum is ....wheat

a) Diploid
b) Hexaploid
c) Tetraploid
d) None

252. Tobacco belongs to the family

a) *Asteraceae*
b) *Poaceae*
c) *Solanaceae*
d) *Brassicaceae*

253. A short duration crop in between two main crops is called as

a) Trap crop
b) Catch crop
c) Cover crop
d) None

254. The leaf area per unit land area is known as

a) LAI
b) LAR
c) CGR
d) RGR

255. The optimum temperature for ideal wheat seed germination is

a) 10-15°C
b) 15-20°C
c) 20-25°C
d) 25-30°C

256. Intercropping is successful if LER is

a) 1 b) >1

c) <1 d) None

257. Silt have particle size range between

a) 2.0-0.2 mm b) 0.2-0.002 mm

c) 0.02-0.002 mm d) <0.002 mm

258. Gypsum is commonly used for reclamation for

a) Saline soil b) Alkali soil

c) Acidic soil d) Waterlogged soil

259. The zinc deficiency in rice causes the

a) Khaira b) White bud

c) Whiptail d) Akiochi

260 Which one is crop rotation

a) Rice–Rice-Rice b) Rice-Green Gram

c) Rice–Wheat-Mustard-Rice d) None

261. Effective Weedicide for the control of *Phalaris minor*

a) Isoproturon b) 2, 4-D

c) Fluchloralin d) None

262. Most commonly used herbicide in pulses is

a) Basalin b) Stam F-34

c) Machete d) Tok E25

263. In zero tillage the surface soil layer have

a) High infiltration of water b) High bulk density

c) Low bulk density d) None

264. The instrument used to measure humidity is

a) Hygrometer b) Hydrometer

c) Pyranometer d) None

265. .Major loss of water used by plants is

a) Evaporation b) Evapotranspiration

c) Metabolic activity d) None

266. Which one of the following is a selective herbicide?
a) Atrazine
b) Diquat
c) Paraquat
d) Glyphosate

267. Sugarcane seed setts should essentially have
a) 4 buds
b) 3 buds
c) 2 buds
d) 1 bud

268. Which one of the following is a biofertilizer?
a) MOP
b) SSP
c) BGA
d) SOP

269. Which of the following organism is responsible for the decomposition of organic matter
a) Algae
b) Fungi
c) Bacteria
d) Microorganisms

270. The losses of ammonia usually becomes greater with
a) Increase in soil pH
b) Decrease in soil pH
c) Increase in soil temperature
d) Increase in soil calcium

271. To prevent denitrification losses of N_2 in paddy field, Nitrogen should be applied in the
a) Oxidized zone
b) Reduced zone
c) Hydronized zone
d) Calcified zone

272. Parallel cropping is a system of copping to make the maximum use of
a) Solar energy
b) Water resources
c) Soil resources
d) All of the above

273. Which one is beneficial element for plants?
a) Mo
b) Cl
c) V
d) Ni

274. Magnesium is taken by plants in the form of
a) Mg^{++}
b) MgO
c) $MgCO_3$
d) None

275. Which one is not a secondary element?
a) Mn
b) Mg
c) S
d) Ca

276. The recommended depth of sowing of wheat is

a) 5 cm b) 3 cm

c) 9 cm d) 7.5 cm

277. *Chenopodium album* in the field of wheat is an example of

a) Absolute weed b) Relative weed

c) Rogue weed d) None

278. 13. The intensity of crop rotation may be indicated by

a) Number of crops grown in rotation x100/duration of the rotation

b) Number of crops grown in rotation/duration of the rotation x100

c) Number of crops grown in rotation/duration of the rotation

d) None

279. Which of the following crop has epigeal type of germination?

a) Soybean b) Rice

c) Pea d) Wheat

280. Which one is trap crop?

a) Bajra b) Sanai

c) Brinjal d) Merigold

281. Which one is a indicator crop of Boron

a) Bajra b) Sunflower

c) Jowar d) Onion

282. Available form of phosphorus ion is

a) Phosphoric acid ($H_2PO_4^{-2}$) b) Phosphorus chloride

c) Phosphorus sulphate d) Phosphorus nitrate

283. The bacteria considered most important to bring out the conversion of NH_4+ to NO_2^-

a) *Nitrosomonas* b) *Nitrobactor*

c) *Nitrogenase* d) *Rhizobium*

284. Wheat is a

a) Long day plant b) Short day plant

c) Day neutral plant d) None of the above

285. Which one is not a cover crop?
a) Moong b) Maize
c) Urd c) Lobia

286. Nutrients absorbed by the plants are carried upward through
a) Xylem b) Phloem
c) Both d) None

287. Which of the following method is most useful to increase the availability of soil applied P_2O_5
a) Broadcast b) Band placement
c) Broadcast and incorporation d) None

288. The bacteria responsible for fixation of nitrogen in soybean is
a) *Rhizobium meliloti* b) *Bradyrhizobium japonicum*
c) *Rhizobium trifolii* d) *Rhizobium lupini*

289. Which one is not a C_3 plant?
a) Wheat b) Rice
c) Barley d) Maize

290. Karnal bunt in wheat is caused by
a) Bacteria b) Fungus
c) Zn deficiency d) Virus

291. Niger is a
a) Pulse crop b) Fibre crop
c) Cereal crop d) Oilseed crop

292. *Vigna mungo* is the botanical name of
a) Moong b) Urd
c) Cowpea d) Arhar

293. Soils having high pH are deficient in
a) Zn b) Mo
c) S d) B

294. Botanical name of sweet corn is
a) *Zea mays saccharata*
c) *Zea mays amylacea*
b) *Zea mays indurata* (Flint)- India
d) Zea mays indentata (Dent)- USA

295. Seed plot technique is practiced in

a) Potato | b) Cotton
c) Rice | d) Tobacco

296. *Azotobactor* fixes atmospheric nitrogen

a) Symbiotically | b) Non-symbiotically
c) Both of the above | d) None of these

297. Rock phosphate is used in

a) Alkaline soil | b) Saline soil
c) Acidic soil | d) Normal soil

298. Water use efficiency is maximum with

a) Furrow irrigation | b) Flooding
c) Border irrigation | d) Drip irrigation

299. The calcium deficiency symptom initially appears on

a) Lowest leaves | b) Top/Apical bud
c) Middle leaves | d) None

300. Pulse crop which does not fix the atmospheric nitrogen significantly

a) Arhar | b) Pea
c) Soybean | d) Rajmash

301. Which type of algae fixes the atmospheric nitrogen into the soil?

a) Blue | b) Blue-green
c) Red | d) None

302. Wind speed is measured with

a) Wind vane | b) Anemometer
c) Barometer | d) None

303. *Adsali* sugarcane is planted in

a) Oct-Nov | b) June-July
c) Feb-March | d) March-April

304. Maturity in sugarcane can easily be determined by hand refractmeters when its reading is

a) 10-15 | b) 17-18
c) 20-22 | d) >22

305. The main objective of puddling in paddy is
   a) Destroy insect-pest
   b) Create impermeable layer on the soil
   c) Mix fertilizer in the soil
   d) Increase of soil air

306. **Awarodhi** is a wilt resistant variety of
   a) Gram
   b) Arhar
   c) Pea
   d) Sugercane

307. Which chemical is used for treatment of sugarcane setts?
   a) Diathane M-45
   b) Dimecron
   c) Aretan
   d) Agrosan GN

308. *Diara* lands are
   a) Area flooded by sea
   b) Area on both sides of canal
   c) Area created by thunder storms
   d) Area located on either sides of the river which get flooded every year

309. Which one of the following development has helped in the success of zero tillage?
   a) Insecticides
   b) Herbicides
   c) Fungicides
   d) Soil amendments

310 Water requirement of a crop is related to its
   a) Evapotranspiration
   b) Dry matter content
   c) Soil mineralogy
   d) Potential absorption coefficient

311. Delinting of cotton seed helps in
   a) Grading of seeds
   b) Killing of hibernating boll worm larvae and disease pathogens
   c) Quicker germination
   d) All of these

312. Montmorillonite, Kaolinite and Illite are
   a) Kinds of soil microbes
   b) Clay minerals
   c) Kinds of protozoa present in soil
   d) Kinds of soil in USA

313. SSP contains .... % sulphur

a) 6 b) 12

c) 18 d) 24

314. Density of water is maximum at

a) -4°C b) 4°C

c) 1°C d) -1°C

315. Most important role in the formation of chlorophyll is performed by

a) Na b) K

c) Mg d) Ca

316. Recommended hills per square metre for paddy transplanting in Usar land ranges between

a) 40-50 b) 30-40

c) 60-70 d) 70-80

317. Defuzzing of seed before sowing is the important practice for the seed of

a) Cotton b) Potato

c) Gram d) Tomato

318. Hydrolysis of urea in soil results in

a) Uric acid b) Ammonium carbonate

c) Ammonium carbamate d) Amides

319. Nozzle suitable for spraying of herbicides is

a) Flat fan b) Hollow cone

c) Solid cone d) None of the above

320. Electrical conductivity of saturation extract of saline soils in dS/m is

a) >4.0 b) 2.0

c) 2.0-3.0 d) 3.5

321. Ratio of standard deviation to the population mean is

a) Coefficient of variation b) Variance

c) Standard error d) None of these

322. Dark reaction takes place in

a) Grana b) Stroma

c) Cristae d) DNA

323. Ratoon stunting of sugarcane is caused by
a) Virus
b) Fungus
c) Bacteria
d) Mycoplasma like organism

324. Sodic soils are characterized by
a) High soluble salt concentration and low pH
b) High soluble salt concentration
c) Low pH and high exchangeable sodium
d) High exchangeable sodium and high pH

325. Which of the following insecticide may be recommended for the control of termite?
a) Dimethoate
b) Nimbicidine
c) Methyl O demeton
d) Chloropyriphos

326. Pink boll worm is a pest of
a) Mustard
b) Cotton
c) Gram
d) Ladies finger

327. In Universal soil loss equation (RKLSCP) 'K' stands for
a) Rainfall and runoff factor
b) Topographic factor
c) soil erodibility factor
d) Soil cover and management factor

328. V shape & yellowing of lower leaves due to deficiency of
a) N
b) P
c) K
d) S

329. Burning of leaf margin is due to deficiency of
a) N
b) P
c) K
d) S

330. Potato is an indicator for which nutrient
a) N
b) P
c) B
d) K

331. Which N-fertilizer is most suited for tea
a) Ammonium sulphate
b) Ammonium nitrate
c) Ammonium sulphate nitrate
d) Urea

332. Nicotine formed in which part of tobacco plant

a) Root
b) Leaf peduncle
c) Leaf
d) Seed

333. Ratio of organic carbon to organic matter is

a) 0.7-1.0
b) 1.7-1.0
c) 1-2
d) 1.5-2.5

334. Pointed gourd is

a) Monocious
b) Dioecious
c) Both
d) None

335. Which water potential is not in soil

a) Turgor potential
b) Gravitational potential
c) Matric potential
d) Osmotic potential

336. In FIRB* irrigation system there is increase efficiency of

a) Water
b) Fertilizer
c) Both a & b
d) Plant

337. Neutralizing value sequence

a) $Ca(OH)_2$>$CaMg(OH)_2$>$CaCO_3$>CaO
b) $CaMg(OH)_2$>$CaCO_3$> CaO>$Ca(OH)_2$
c) CaO>$Ca(OH)_2$>$CaMg(OH)_2$>$CaCO_3$
d) $CaCO_3$>CaO> $CaMg(OH)_2$>$Ca(OH)_2$

338. Cropping intensity of 166% is of sequence

a) 5 crops for 3 years
b) 4 crops for 3 years
c) 5 crops for 4 years
d) 4 crops for 5 years

339. Boron toxicity is controlled by

a) Lime
b) Gypsum
c) Leaching
d) All

340. Illoxan is used for control of

a) Wild oat
b) Phalaris minor
c) Chenopodium album
d) All

341. Pop corn

a) *Zea mays* indurate
b) *Zea mays* everta
c) *Zea mays* indentata
d) All

342. Soil A with pH 5.0 & soil B pH 7.0 how many times low acidity

a) 10
b) 100
c) 1000
d) 10000

343. If 1000 litres solution contains 2000 ppm chemical how much chemical is there in gram?

a) 20 g
b) 200 g
c) 20.2 g
d) 2000 g

344. The plants meet their carbon requirements by absorbing

a) Soil minerals
b) Soil organic matter
c) Carbon monoxide from atmosphere
d) Carbon dioxide from atmosphere

345. Bending of plants towards light is called

a) Hydrotropism
b) Geotropism
c) Phototropism
d) Chemotropism

346. Nitrogen is a major nutrient because

a) More availability in soil
b) Constituent of protein
c) It is available in atmosphere more
d) It is required by plants more

347. Local control helps in

a) Reducing the experimental error
b) Homogeneity of experimental units
c) Both
d) None

348. Arithmetic mean is most commonly used because

a) Based on all the observations
b) Largest and smallest observations
c) Located in the middle of the series
d) None of these

349. Extension education is .... process

a) Only teaching
b) Only learning
c) Teaching and learning
d) None of these

350. Weed competition in rice is more in

a) Transplanted crop
b) Direct seeded crop
c) Flooded crop
d) Crop sown in rows

351. The artificial application of water to soil for the purpose of supplying the moisture essential for plant growth is termed as

a) Irrigation
b) Drainage
c) Irrigation requirement
d) Water requirement

352. The average annual precipitation for the country as a whole is.

a) 575 mm
b) 1200 mm
c) 1000 mm
d) 1100 mm

353. In the Negeb desert, southwest of the dead sea, which form of precipitation is largely responsible for summer growth of grapes.

a) Rainfall
b) Fog
c) Dew
d) Hailstorms

354. The water beneath the soil surface where voids in the soil are substantially filled with water is called

a) Pheratic water
b) Surface water
c) Aquifer
d) Ground water

355. Which one of the following countries has 100 per cent cultivated area under irrigation

a) Egypt
b) India
c) USA
d) Japan

356. The devices and methods used to control silting of reservoirs include.

a) Settling basins and by – pass channels
b) Vegetated streams and venting density currents
c) Flood sluicing, dredging, draining and flushing
d) All of the above

357. ————— are water-loving plants growing along stream courses and on wet soils having high water tables where an abundant supply of water is available

a) Phreatophytes b) Hydrophytes

c) Aquatic plants d) None of these

358. The science of artificial rain – making is known as

a) Forced raining b) Cloud seeding

c) Cloud bursting None of thesed)

359. The chemical used for seeding of cold clouds is

a) NaCl b) AgI

c) Both a & b d) None

360 The chemical used for seeding of warm clouds is.

a) NaCl b) Urea + concentrated NH_4NO_3

c) Both a & b d) None

361. The amount of electrical energy required to remove the salt from 1 M^3 of sea water is

a) 0.5 Kilowatt hours b) 0.6 Kilowatt hours

c) 0.7 Kilowatt hours d) 0.8 Kilowatt hours

362. The soils having excess soluble salts are designated as.

a) Alkali soils b) Saline – alkali soils

c) Saline soils d) All of the above

363. Which one of the following soil textural class has the highest infiltration and permeability

a) Sandy b) Sandy loam

c) Clay loam d) Clay

364. The soil textural class having the highest bulk density is.

a) Clay b) Clay loam

c) Sandy loam d) Sandy

365. The available water range of loamy soil is

a) 4-6% b) 6-10%

c) 10-14% d) 14-18%

366. Tensiometers are most useful in ———— soils.

a) Sandy b) Clay

c) Vertisols d) All

367. The leaching requirement increases at least ———— percent for each 1000 ppm of dissolved solids in the inigation water

a) 1 b) 10

c) 100 d) 1000

368. The plants in saline soils are adversely affected by

a) High concentration of salts b) Poor soil physical conditions

c) Both d) None

369. The boron content of inigation water suitable for inigation is

a) <2 ppm b) >2 ppm

c) <2% d) >2%

370. The crop having the highest tolerance to boron is

a) Sugarbeet b) Sunflower

c) Alfalfa d) Apple

371. The major source of heat used for transpiration is

a) Terrestrial energy b) Radiant energy

c) Both d) None

372. The total pore space of the soil less the moistme content at field capacity, both expressed as volume percentages of the total soil volume is defined as

a) Specific yield b) Mixture equivalent

c) Water yield d) Total yield

373. Blancey and criddle developed a simplified formula for measurement of consumptive use based on

a) *Temperature and day – time hours*

b) RH, mind velocity, sunshine hours and Temperature

c) Temperature and latitude

d) All of the above

374. The shallow–rooted crops will require ———— irrigation than deep–rooted crops
   a) More frequent b) Less frequent
   c) Deep d) Shallow

375. Plants with restricted root systems show symptoms of
   a) Low fertility b) Drought
   c) Deformed tap roots d) All of the above

376. The water – conveyance efficiency (Ec) can be stated as.
   a) Wf × 100 / Wd b) Ws × 100 / Wf
   c) Ws × 100 / Wn d) 1 – y × 100 / d

377. Which efficiency will provide the measure for comparison between systems whether it be sprinkler compared to sprinkler, surface compared to surface or sprinkler compared to surface irrigation.
   a) Ea b) Ed
   c) Ec d) Es

378. The method of sprinkler inigation was started about ———— AD
   a) 1800 b) 1900
   c) 1700 d) 1950

379. The method of applying water to the surface of the soil in the form of a spray, somewhat as ordinary rain, is known as
   a) Drip irrigation b) Bubbler irrigation
   c) Sprinkler irrigation d) Foggers

380. Which method of irrigation can provide protection against frost
   a) Flooding b) Cheek – basin
   c) Drip irrigation d) Sprinkler irrigation

381. The benefits of trickle irrigation include
   a) Minimization of insect, diseare and fungus problems
   b) Less soil crusting
   c) Water saving (30-50 %) compared to other methods
   d) All of the above

382. The method of irrigation suitable of row crops is
   a) Flood b) Check basin
   c) Furrow d) Drip

383. The implements used for land – leveling include

a) Scrapers b) Soil movers

c) Both d) None

384. 1 hectare metre is equal to

a) 10 m^3 b) 100 m^3

c) 1000 m^3 d) 10,000 m^3

385. The discharge (Q) in l/ second of a submerged orifice can be expressed (A = Area, cm2 ; H = head, cm L = length, cm)

a) 0.027 A"h b) 0.0184 $LH^{3/2}$

c) 0.0186 $LH^{3/2}$ d) 0.0138 $H^{5/2}$

386. The contribution of south – west monsoon in total precipitation of the country is around

a) 400 m ha m b) 100 m ha m

c) 300 m ha m d) 200 m ha m

387. The water movement in the soil comprises of

a) Infiltration b) Redistribution

c) With drawl d) All of the above

388. Which type of water movement in the soil is more important from crop production point of view

a) Saturated flow b) Unsaturated flow

c) Vapour movement d) All of the above

389. The volume of water movement in the soil is highest in

a) Saturated flow b) Unsaturated flow

c) Vapour movement d) All

390. The evapotrauspiration constitutes nearly —— per cent of total water uptake

a) 90 b) 80

c) 99 d) 85

391. Water is retained in the soil by

a) Adhesion b) Cohesion

c) Both of the above d) None of these

392. The energy status of water at saturation is

a) 0 bar
b) 100 bar
c) 50 bar
d) 75 bar

393. The amount of available water present in clay soil (cm/m depth) is

a) 8
b) 12
c) 17
d) 23

394. The portion of water potential that results from the solutes present in the soil is referred to as

a) Matric potential
b) Osmotic potential
c) Gravitational potential
d) All of the aboved)

395. Tensiometers are also called

a) Irrometers
b) Water meters
c) Current meters
d) None of these

396. The shape of the soil moisture characteristic cmue in sandy soils is

a) Straight line
b) L – shaped
c) S – shaped
d) All

397. The relationship between dry matter of crop and ET in cereals is

a) Linear
b) Quadratic
c) No relationship
d) Sigmoidal

398. Which is the most important factor that decides the rate of ET.

a) Soil conditions
b) Climatic conditions
c) Crop characteristics
d) None

399. Crop coefficient (Kc) is the ratio between

a) ETc /ETo
b) ETo / ETc
c) ETa / E
d) None of these

400. The 'Drum culture technique' is used for measuring components of WR of

a) Wheat
b) Rice
c) Sugercane
d) Jute

401. The irrigation requirement (IR) of a crop can be expressed a

a) WR + ER + S
b) WR – (ER + S)
c) IR + ER + S
d) None of these

402. When rainfall is of low intensity and light and ground cover is full, the effective rainfall is close to

a) 0% b) 50%

c) 75% d) 100%

403. Calculate the amount of water for each irrigation for scheduling irrigation at 0.8 IW/CPE ratio with 10 cm of CPE.

a) 5 cm b) 8 cm

c) 2 cm d) 11 cm

404. The most critical stage of irrigation in rice is

a) Flowering b) Booting

c) Panicle initiation d) All of above

405. Which is the most common method among the surface method of irrigation

a) Flooding b) Check basin

c) Basin method d) Furrow method

406. The intermittent application of water to the field surface under gravity flow resulting in a series of 'on' and 'off' nroder of constant or variable time spans is defined as

a) Cablegation b) Surge irrigation

c) Sprinkler irrigation d) Trickle irrigation

407. Which one of the following countries has the largest area under drip irrigation

a) USA b) Israel

c) Australlia d) India

408. The 'Typhoon system' of irrigation is useful for

a) Sugarcane b) Paddy

c) Citrus d) Wheat

409. The field water use efficiency is defined as

a) Y / ET b) Y / WR

c) Y / CU d) Y / IR

410 Which one of the following has the highest WUE

a) Rice b) Maize

c) Wheat d) Finger millet

411. The susceptible crops for water logging include
a) Tobacco
b) Tomato
c) Pulses
d) All

412. The injury to plants due to water-logging is severe under——— conditions
a) Warm and summy weather
b) Cool and cloudy weather
c) Warm and cloudy weather
d) Cool and sunny weather

413. Water – logging causes injury to the plants due to
a) Low O_2 content
b) Toxins accumulation
c) Both of the above
d) None of these

414. Which one of the following is used as an indicator plant to irrigation in sugarcane
a) Mango
b) Potato
c) Banana
d) Sunflower

415. The drip irrigation system was introduced in India during early
a) 1950's
b) 1960's
c) 1970's
d) 1980's

416. A soil core of 14 cm diameter and 10 cm height weighs 2772 g of fresh weight and over dry weight is 2310 g. Find out bulk density
a) 1.5 Mg / m^3
b) 1.6 Mg / m^3
c) 1.7 Mg / m^3
d) 1.8 Mg / m^3

417. If a soil has Dp = 2.65 Mg / m^3, Db = 1.40 Mg / m^3 and field capacity = 25 % on oven – dry weight basis, calculate capillary and non – capillary porosity
a) 35 % and 12.17 % respectively
b) 47.17 and 12.17 % , respectively
c) 12.17 % and 50.0 % , respectively None of thesed)

418. Which method of soil moisture measurement is useful scheduling irrigation to frequently irrigated crops
a) Electrical resistance
b) Tensiometers
c) Neutron moisture meter
d) Thermocouple Psychrometer

419. The best method for quantitative estimation of changes in soil moisture *in situ* is
a) Tensiometer
b) Electrical resistance
c) Neutron moisture
d) Pressure plate apparatus

420. The standard evaporimeter is

a) Piche evaporimeter
b) USWB class A pass evaporimeter
c) Sunken screen evaporimeter
d) None of the above

421. The sunken screen evaporimeter was developed at IARI by

a) Sharma and Dastane
b) Parihar *et al.*
c) S. B. Ray and K. K. sinha
d) None of these

422. Wheat crop requires 40 cm of irrigation water during 120 days irrigating period. How much land can be irrigated with a flow of 20 l / second for 12 hours a day.

a) 25.9 ha
b) 28.9 ha
c) 2.59 ha
d) 2.89 ha

423. A 24 – litre capacity bucket is filled in 10 seconds by the discharge from a Persian wheel. What is the rate of flow?

a) 0.24 lit / sec
b) 2.4 lit / sec
c) 144 l / min
d) Both 2 and 3

424. Which device is commonly used to measure discharge from small and medium sized streams

a) Parshall flume
b) V – notch weir
c) Orifice
d) All

425. After how many days will you apply water to soil if FC of the soil = 28 %; PWP = 13 %; Db = 1.3 Mg / m^3; effective root zone depth = 70 cm and daily Cu of water for the crop is 12 mm; Assume additional data if necessary

a) 11 days
b) 13 days
c) 15 days
d) 17 days

426. Suppose the grain yield of wheat is 45 qha^{-1} and ET = 40 cm crop water use efficiency will be

a) 112.5 Kg / ha – cm
b) 112.5 q / ha – cm
c) 11.25 Kg / ha – cm
d) None of these

427. Which one of the following is a water conveyance structure

a) Flumes
b) Culverts
c) Inverted siphon
d) All

428. If the amount of water required to refill the soil profile is 5 cm and the irrigation efficiency is 60 per cent, the total depth of water to be applied would be

a) 8.33 cm
b) 1.20 cm
c) 6.33 cm
d) 3.20 cm

429. The perforated pipe sprinkler system operates at a pressure of

a) 1.4 kg cm^{-2} and less
b) 2 $kgcm^{-2}$ and more
c) Both of these
d) None of these

430. The vertical distance above a water table along which moisture content varies from full saturation to field capacity is called

a) Capillary fringe
b) Aquifer
c) Perched water table
d) None of these

431. The sunken screen evaporimseter was developed at IARI in 1968 by.

a) Dastane and Sharma
b) RBL Bhardwaj
c) Krishnan and singh
d) Parihar et al

432. The soil developed under conditions of poor drainage resulting into grey or motlle colour due to reduction if iron and other elements are

a) Gley soil
b) Marsh soil
c) Diara lands
d) Kachhar lands

433. Which of the following is a stomatal closing type of antitranspirant

a) PMA
b) Adol 52
c) OED green
d) All

434. The portion of precipitation which is lost through surface flow from an area without entering the soil is termed as.

a) Runoff
b) Leaching
c) Percolation
d) All of the above

435. The concept of minimum tillage was started in

a) India
b) USA
c) UK
d) Australlia

436. The matric potential of a saturated soil is.

a) 0
b) -1
c) 1
d) None

437. The water requirement of maize ranges in between

a) 20 – 40 cm b) 40 – 60 cm

c) 60 – 80 cm d) 80 –100 cm

438. The lack of resistance of a soil to soil erosion is referred to as

a) Erosivity b) Erodibility

c) Erosion accelerated d) Erosion

439. According to NARP, the country has been divided into ———— Agro–climatic zones

a) 20 b) 15

c) 10 d) 127

440. The dry land soils are ——— in organic matter content

a) Low b) Medium

c) High d) Very high

441. The lower the O.M. content of the soil, the —— will be N and P content of soil

a) Low b) Medium

c) High d) Very high

442. The dryland soils are deficient in N but are slightly high in

a) P b) K

c) Fe d) Zn

443. Which crop can be successfully grown in heavy black dry lands during Rabi season

a) Sunflower b) Castor

c) Safflower d) Rapeseed & mustard

444. How much of the water absorbed by the plants is lost in transpiration

a) 70% b) 80%

c) 90% d) 99%

445. Kaolin reduces the transpiration loss of water by

a) Increase of leaf reflectance

b) Reduction of leaf temperature

c) Reduction of vapour pressure gradiart

d) All of the above

446. Which chemical is used as an antitranspirant in drylands
a) PMA
b) Cycocel
c) Kaolin
d) All of these

447. Which are used as anti=transpirants for closure of stomata
a) PMA
b) Atrazine
c) ABA
d) All of these

448. The dryland agriculture which covers about 63 % of the net soun area contributes to the foodgrain production in the country to the true of
a) 39%
b) 41%
c) 43%
d) 45%

449. The water harvesting method in which the harvested water is stored in the soil profile itself is called as
a) Run oft concentration
b) Run–off farming
c) In–situ water harvesting
d) All

450. The pit and trench method of water harvesting for vegetables and fruit crops was developed by
a) CAZRI
b) ICRISAT
c) IARI
d) ICARDA

451. The water harvesting technique in which run off is stored in a reservoir for
a) Run off recycling
b) Run off collection
c) Catchment water harvesting
d) All of the above

452. Which of the following chemicals / materials have been extensively used to induce the run off
a) NaCl
b) Bentonite
c) Asphalt
d) All of the above

453. Which of the following is used to retard water loss from the ponds
a) Cetyl alcohol
b) Ethyl alcohol
c) Methyl alcohol
d) All of the above

454. The traditional water harvesting techniques includes.
a) Baories and Talabs
b) Khadins and Tanka's
c) Talai and Samand
d) All of the above

455. The main factors responsible for soil erosion in India are.

a) Excessive deforestation b) Overgrazing

c) Faulty agricultural practices d) All of the above

456. The natural or normal erosion is also known as

a) Geolgical erosion b) Acceleraed erosion

c) Abnorural erosion d) All of the above

457. The water – use efficiency of intercropping is ——— than sole crops

a) Higher b) Lower

c) Equal d) None of the above

458. The highest crude protein (%) content is in seeds of

a) *Prosopis juliflora* b) *Albizia lebeck*

c) *Acacia nilotica* d) *Prosopis cinecaria*

459. When agricultural crops are grown in combination with tree species, the system is known as

a) Silvipasture b) Agrisilviculture

c) Agri-silvipasture d) Agrisilviaquaculture

460 Which of the following plant species are know to fix atmospheric nitrogen

a) *Acacia* b) *Leucaena*

c) *Casuarina* d) All of the above

461. In India, shifting cultivation is practiced in.

a) NEH region b) Jharkhand

c) AP d) All of the above

462. The 'Taungya' is a ———— word, consisting of 'Taung' means hill and 'Ya' means cultivation i.e. cultivation in the hills

a) Assamese b) Arabic

c) Burmese d) Hindi

463. The *Taungya* system was introduced to India in 1856 by

a) Blanford b) Brandis

c) Kang and Wilson d) King

464. The inherent capability of an organism to multiply over a given period of time is called

a) Biotic potential b) Brecding potential

c) Climatic potential d) None of these

465. The instrument 'Dendrometer' is used to measure
   a) Tree hight b) Trunk diameter
   c) Bole diameter d) None of these

466. The epiphytes grow on other plants mainly for
   a) Water absorption b) Mechanical support
   c) Nutrients d) All of the above

467. Hypsometer is used to measure ———
   a) Trunk diameter b) Bole diameter
   c) Tree height d) All of the above

468. The business of cutting or getting timber or from the forest for lumber
   a) Lumbering b) Logging
   c) Lodging d) Looping

469. The study of the life history and general characteristics of trees and forests is termed as
   a) Silviculture b) Silvics
   c) Shelterbelts d) Singling

470. The proportion of area under various crops at a point of time in a unit area is described as
   a) Cropping pattern b) Cropping system
   c) Crop rotation d) Crop diversification

471. At present, the main objective of intercropping is
   a) Insurance against crop failure b) Higher productivity per unit area
   c) Stability in production d) Efficient utilization of resources

472. The non-host crop planted to make nematodes waste their infection potential are
   a) Decoy crops b) Trap crops
   c) Cover crops d) Guard crops

473. The intensive cropping is possible with the introduction of.
   a) Short duration photo sensitive varieties
   b) Short duration photo insensitive varieties
   c) Long duration photo insensitive varieties
   d) None of these

474. The most efficient crop for N and P utilization is.

a) Jute b) Summer rice

c) Potato d) Maize

475. The most efficient crop for K utilization is

a) Groundnut b) Maize

c) Jute d) Potato

476. Which one of the following is the most intensive grazing system

a) Continuous grazing b) Rotational grazing

c) Rational grazing d) None of these

477. The separation of grain or seed from chaff is known as

a) Threshing b) Harvesting

c) Drying d) Winnowing

478. Crop storage means

a) Storage in open b) Storage in godowns

c) Storage in underground birs d) All of the above

479. The growing of which crop reduces the population of *pratylenchus* eel worms

a) Tomato b) Castor

c) Marigold d) Gingelly

480. The farmers practice monocropping and monoculture due to

a) Climatic conditions b) Socio-economic conditions

c) Specialization of a farmer d) All of the above

481. The mixed cropping as a common practice in most of———— tracts of India

a) Irrigated b) Dryland

c) Water-logged d) None of these

482. Who is considered as the father of tillage

a) Jethro Tull b) G.B. Triplett

c) Glauber d) P.C. Raheja

483. Which one of the following is a multipurpose plough.

a) M.B. Plough b) Disc plough

c) Country plough d) All of the above

484. One centimeter of surface soil over one hectare of land weighs about ———— tones

a) 100 b) 150

c) 200 d) 250

485. According to CRIDA, ploughing of——— cm depth is deep ploughing

a) 5-6 b) 15-20

c) 25-30 d) 30-35

486. The tillage operations carried out through out the year are

a) Year-round tillage b) Minimum tillage

c) Zero-tillage d) None of these

487. The 'after cultivation' practice includes

a) Drilling of fertilizers b) Earthing up

c) Blind hoeing d) All of the above

488. Which plough is the smallest of the wooden ploughs

a) Black soil plough b) Dryland plough

c) Wetland plough d) None of these

489. Which of the following is not a secondary tillage implement

a) Cultivator b) Harrow

c) Plough d) Planker

490. When the primary tillage is completely avoided and secondary tillage is restricted to seedbed preparation in the row zone only then it is defined as

a) Zero-tillage b) Minimum tillage

c) Reduced tillage d) Conventional tillage

491. A single tillage operation over a hectare with a country plough requires —— Km walk on rough land.

a) 10 b) 20

c) 30 d) 40

492. The optimum depth of sowing for most of the field crops

a) 3-5 cm b) 1-2 cm

c) 5-7 cm d) 7-9 cm

493. As a thumb rule, what would be the optimum age of seedlings for every month of total duration of the crop

a) 1 week b) 2 week
c) 3 week d) 4 week

494. The shape of space available for individual plants is referred to as

a) Plant geometry b) Crop geometry
c) Rectangularity d) None of these

495. The agronomic practice of gap filling is advantageous in

a) Short duration and closely spaced crops
b) Short duration and widely spaced crops
c) Long duration and widely spaced crops
d) None of these

496. The actual site of action for photosynthesis is.

a) Merophyll cells b) Mitochondria
c) Ribosanes d) Bundle sheath cells

497. The germination found in monocot plants is

a) Epigeal b) Hypogeal
c) Pseudohypogeal d) None of these

498. When the seedlings are transplanted, they establish by producing

a) Crown roots b) Nodal roots
c) Prop roots d) Seminal roots

499. The fruit growth in cotton is also called as

a) Grain growth b) Pod growth
c) boll growth d) None of these

500. The optimum LAI for crops with horizontally oriented leaves ranges in between

a) 3-4 b) 6-9
c) 10-13 d) 4-7

501. The idea of wilting coefficient was introduced in 1912 by

a) Briggs and Shantz b) Briggs and Mclane
c) Hains d) Buckingham

502. The properties of humus are in general attributed to the preserves of

a) Fulvic acid b) Humic acid

c) Humin d) Hematomelanic acid

503. The book "Soil conditions and plant Growth" is written by

a) Nyle C. Brady b) M. Alexander

c) E.J. Russell and E.W. Russell d) S.A. Waksman

504. The number of microorganisms and their activities in the soil are the highest in

a) Winter b) Summer

c) Spring d) None of these

505. The process of oxidation involves

a) Addition of O_2 b) Removal of Hydrogen

c) Loss of electrons d) All of the above

506. Which one of the following algae are rarely found in soil.

a) BGA b) Grass-green algae

c) Yellow-green algae d) Diatoms

507. The role of algae in supplying O_2 to the roots of submerged rice plants was investigated in 1913 by

a) W.H. Harrison & P.A.S. Aiyer b) J.C. Gilman

c) L.D. Galloway d) None of these

508. Which genera of soil fungi are known to fix elemental nitrogen?

a) *Phoma* b) *Rhizoctonia*

c) *Boletus* d) All of the above

509. The genera *Nitrobacter* and *Nitrosomonas* of bacteria belong to order

a) Pseudomonadales b) Eubacteriates

c) Myxobacterales d) None of these

510 The scientific name of Egyptian clover is

a) Medicago sativa b) Melilotus parviflora

c) Trifolium alexandrinum d) Ornithopus sativvs

511. The organism responsible for symbiotic nitrogen fixation in *Ornithopus sativus* (Bird's foot) is

a) *Rhizobium lupini* b) R. phaseoli

c) *R. leguminosarrum* d) *R. japonicum*

512. The addition of which nutrient element to soil increases the nitrogen fixing power of the legume

a) Nitrogen b) Phosphorus

c) Potassium d) None of these

513. Who for the first time isolated the bacteria *Nitrosomonas* and *Nitrobacter* responsible for nitrification process in soil?

a) Warington b) Schloesing and Muntz

c) Winogradskey d) Fred, Baldwin and Mccoy

514. Under anaerobic conditions, the decomposition of cellulose is brought about entirely by

a) Algae b) Fungi

c) Actinomycetes d) Bacteria

515. The unproductiveness of acid or alkaline soils is vary often due to the.

a) Lack of available plant nutrients b) Acid or alkaline nature of medium

c) Both of the above d) None of these

516. In highly acid soils, the availability of which nutrients have increased to toxic levess to the plants

a) Al b) Fe

c) Mn d) All of the above

517. The first volume of the Journal "Advances is Agronomy" was published in

a) 1947 b) 1949

c) 1951 d) 1953

518. The most harmful salt to plants in saline and alkaline soils is

a) Na_2SO_4 b) NaCl

c) Na_2CO_3 d) None of these

519. The concentration of CO_2 in soil air is highest under

a) Permanent pastures b) Cultivated soil

c) Fallow lands d) None of these

520. The insufficient soil acration lands to the develop ment of plant diseares such as.

a) Wilt
b) Citeus dieback
c) Both of these
d) None of these

521. A system of soil classification based on the extent of base saturation and the nature of exchangeable bases was proposed by

a) Gedroiz
b) Masbut
c) De' sigmond
d) Glinka

522. When the land was assessed on the basis of the suitability of the soil for certain crops and of its crop producing capacity under local conditions, it is known as.

a) Annewari system
b) Land capability system
c) Crop capacity system
d) None of these

523. The predominant clay mineral in lateritic and lateritic soils is

a) Montmorillonite
b) Illite
c) Halloysite
d) None of these

524. The peaty saline soils are locally known as ———— soils in Kerala.

a) Kallar
b) Kari
c) Khajan
d) Reh

525. The journal "Soils and Fertilizers" is published from.

a) USA
b) UK
c) India
d) Germany

526. In the field, the soil structure is described in terms of.

a) Type
b) Class
c) Grade
d) All

527. Which one of the following is a metamorphic rock

a) Marble
b) Limestone
c) Granite
d) Gabbro

528. The soils formed at higher levels under low moisture conditions

a) Fine-textured
b) Coarse-textured
c) Very fine-textured
d) None of these

529. The good loamy soils may have pose space of——— percent

a) 30 b) 40-50

c) 50-60 d) 80-90

530. The temperature of the soil is influenced by its

a) Colour and composition b) Slope and aspect

c) Water content d) All of the above

531. The specific heat of water is ——— times more than that of the soil particles

a) 5 b) 10

c) 15 d) 20

532. The capillary water is held at a tension ranging from ——— atmospheres

a) 31-10000 b) 1/3 - 31

c) < 1/3 d) None of these

533. The ultimate source of nitrogen for plants is

a) Soil b) Water

c) Atmosphere d) All of these

534. The average nitrogen content (%) of humus is

a) 5 b) 20

c) 35 d) 50

535. Which one of the following land capability class is suitable for mtessive cropping

a) Class I b) Class II

c) Class III d) Class IV

536. Which is the basic unit of soil classification

a) Soil order b) Soil series

c) Great group d) Family

537. In the Gujarat state, the alluvial soils are locally known as

a) *Bhangar* b) *Khadar*

c) *Goradu* d) Usar

538. Which state occupy the largest area under red soils constituting nearly two-thirds of its cultivated area

a) Maharastra b) Tamil Nadu

c) Karnataka d) Kerala

539. The laterites and lateritic soils are formed under ———— climate

a) Tropical humid b) Temperate humid

c) Tropical dry d) Temperate dry

540. Which one of the following is taken into account for reclamation of acidic soils

a) Active acidity b) Total acidity

c) Both of these d) None of these

541. The commonly used "liming factor" is

a) 0.5-1.0 b) 1.0-1.5

c) 1.5-2.0 d) 2.0-2.5

542. The optimum pH range suitable for potato cultivation is

a) 4.0-6.0 b) 5.0-5.5

c) 6.0-7.5 d) 5.0-6.5

543. Which one of the following is the first feature of good soil management

a) Good soil tilth b) Soil fertility

c) Soil water availability d) None of these

544. The desired size of the soil sample for soil testing is (kg)

a) 0.25 b) 0.50

c) 0.75 d) 1.00

545. The organic carbon content of a soil is a measure of

a) Available phosphorus b) Available nitrogen

c) Total nitrogen d) Available potassium

546. The concentration of total soluble salts (mmhos) cm) critical for germination is

a) > 4 b) 2-4

c) < 1 d) 1-2

547. During winter, the northern part of the country gets some rainfall (Mahawat) from

a) Indus b) Ganges

c) Brahamputra d) Luni

548. The major consumer of water resources in the country is.
   a) Domestic water supply   b) Industrial use
   c) Irrigation   d) None of these

549. The major consumer of water resources in the country is
   a) Domestic water supply   b) Industrial use
   c) Irrigation   d) None of these

550. The net area sown (Million ha.) in the country is
   a) 185   b) 142
   c) 328   d) 55

551. The largest state in terms of geographical area is
   a) Madhya Pradesh   b) UP
   c) Maharastra   d) Rajasthan

552. According to 2011 census, the population of the country is
   a) 1027 million   b) 1000 million
   c) 846 million   d) 1210 million

553. The percentage arable land exploited can be expressed as
   a) Net area sown /Potential land × 100
   b) Net area sown / Arable land ×100
   c) Arable land / Net area sown × 100
   d) None of these

554. The per capita arable land is highest in
   a) India   b) Russia
   c) Canada   d) Australlia

555. Which Indian state accounts for over two-fifths of the nation's culturable wasteland
   a) MP   b) Rajasthan
   c) UP   d) Punjab

556. Which state has the highest net area sown in the country
   a) MP   b) UP
   c) Maharastra   d) Punjab

557. How many years, nature requires to build up 2.5 cm of top soil
   a) 500   b) 1000
   c) 1500   d) 2000

558. The movement of soil by the wind in a series of short bounces is called

a) Saltation b) Suspenion

c) Surface creep d) None of these

559. Which type of soils are more susceptible to erosion

a) Coarse-textured b) Fine-textured

c) Medium-textured d) None of these

560 Which of the following is not a form of strip cropping

a) Contom strip-cropping b) Wing strp-cropping

c) Buffer strip cropping d) None of these

561. The width of erosion-resistant and erosion-pemitting crops for slops ranging between 1-2 percent should be ———— and ———————— m

a) 6, 24 b) 9, 45

c) 4.5, 13.5 d) All of these

562. Which one of the following are mechanical measures of erosion control

a) Sub-soiling b) Basin-listing

c) Contour bunding d) All of these

563. The contour bunding is suitable upto a slope of

a) 6% b) 15%

c) 16-31% d) None of these

564. The 'Tehri dam' is situated on confluence of the rivers

a) Bhagirathi and Bhilgavga b) Bhagirathi and Alaknanda

c) Yamuna and Gauges d) None of these

565. The projected foodgrain production during 2001-02 is ——— million tones

a) 196 b) 210

c) 200 d) 206

566. The present Director – General of ICAR and secretary to DARE is

a) Dr. R.S. Paroda b) Dr. Panjab Singh

c) Dr. R.B. Singh d) Dr. P.K. Singh

567. Water management practices include

a) Irrigation b) Drainage

c) Both of the above d) None of these

568. The available soil water – holding capacity increases mainly with

a) Fineness of texture b) Organic matter content
c) Both of these d) None of these

569. The equal and uniform water availability concept over the entire available range (FC to PWP) holds true for

a) Perennial species b) Seasonal field crops
c) Forage crops d) All of these

570. The irrigation efficiency depends upon

a) Soil texture b) Management practices
c) Stream size d) All of these

571. If a soil has 20 per cent moisture on the over dry basis and weighs 1.5 g per cubic centimeter ; calculate moisture content on volume basis

a) 13.33% b) 30.00%
c) 20.00% d) None of these

572. 1 cubic metre of water is equal to

a) 35.32 cft b) 220 gallons
c) 1000 litres d) All of these

573. The major portion o water applied to the rice crops is lost through

a) Leaching b) Deep percolation
c) Evaportation d) All of these

574. The excess of which plant nutrient causes lodging of cereals

a) N b) P
c) Ca d) K

575. The intraveinal chlorosis and 'firing' along the edges of maize leaues is caused due to deficiency of

a) N b) P
c) K d) Fe

576. The removal of K_2O from soil is highest for

a) Sugarcane b) Groundnut
c) Sorghum d) Wheat

577. The use of manures and fertilizers is ———— for each other

a) Supplementary b) Competitive
c) Complementary d) None of these

578. The most commonly used organic manure in India is

a) FYM b) Compost

c) Green manuring d) Oilcakes

579. The folding 7000 sheep for one night is said to add the equivalent of—— tones of cattle dung

a) 5 b) 10

c) 15 d) 20

580. A leguminous crop producing 8 to 25 tones of green matter per hectare will add about ——— kg of nitrogen when ploughed under

a) 30-60 b) 60-90

c) 90-120 d) 120-150

581. The nitrogen content of Ammonium sulphate nitrate is in

a) 25 % NO_3^- + 75 % NH_4^+ form b) 50 % NO_3^- + 50 % NH_4^+ form

b) 75 % NO_3^- + 25 % NH_4^+ form d) 50 % NO_3^- + 50 % Amide form

582. The NPK content (%) of ammophos is

a) 20-16-0 b) 16-20-0

c) 20-0-16 d) 0-16-20

583. 1 hundred weight (cwt) is equal to

a) 112 lbs b) 122 lbs

c) 132 lbs d) 142 lbs

584. 1 wheat bustle is equal to

a) 42 lbs b) 60 lbs

c) 56 lbs d) None of these

585. The simple perennial weeds multiply only through

a) Seed b) Stolon

c) Rhizomes d) None of these

586. In drylands, the competitions between weeds and crops is largely for

a) Nutrients b) Water

c) Both of the above d) None of these

587. Which of the following is not a contact herbicide

a) Propanil b) Dicryl

c) H_2SO_4 d) Carbyne

588. *Echiochloa colonum* belongs to family

a) Gramineae b) Euporbiaceae

c) Leguminosae d) None of these

589. Which one of the following is a leguminous plant/weed

a) *Meliotus* b) *Tephrosia*

c) *Prosopis* d) All of these

590. *Lantana camara* belongs to family

a) Verbenaceae b) Typhaceae

c) Scrophulariaceae d) Papaveraceae

591. 2,3,6 – trichlorophenyl acetic acid is the chemical name of.

a) Dequat b) Fenac

c) Dalapon d) 2,4 – DES

592. One-seeded indehiscent fruit in which the pericarp is firmly aduate to the seed is called

a) Capsule b) Drupe

c) Achene d) Siliqua

593. The absorbing organs often root-like of some parasitic plants, usually having the anatomy of a stem rather than of a root is termed as.

a) Haustoria b) Stipules

c) Adnate d) None of these

594. Which is the most effective way of controlling air-borne diseases

a) Use of soil disinfectants b) Chemical application on foliage

c) Hot-water treatment d) Solar treatment

595. In northem hill zone, the sowing of wheat is done in

a) October b) November

c) December d) January

596. The optimum time of sowing for short duration varieties of wheat is

a) First fortnight of November b) Second fortnight of November

c) First fortnight of December d) October

597. The optimum row spacing (cm) for irrigated , timely sown wheat is

a) 20.0 b) 22.5

c) 25.0 d) 17.5

598. The wheat variety which is time-tested and remained in cultivation in Australia for several decades is

a) Hope b) Atal

c) C-306 d) Shilaza

599. The rice varieties grown in Indonesia belong to the subspecies or race

a) Indica *b) Japonica*

c) Javanica d) All of these

600. Rice can grow normally in soils up to ———— dsm^{-1} conductivity

a) 2.0 b) 4.0

c) 6.0 d) 8.0

601. Norin 8 and Norin 18 are the varieties of

a) Wheat b) Rice

c) Barley d) Maize

602. Which one of the following crop needs len water and is more tolerant to saline and alkali condition than other winter cereals

a) Barley b) Wheat

c) Oats d) None of these

603. The excess of which plant nutrient affects the malting and brewing quality of barley grain cdvasely

a) P b) K

c) S d) N

604. The first pigconpea hybrid ICPH-8 was released from

a) CAZRI b) IARI

c) IIPR d) ICRISAT

605. The groundnut productivity ($Kgha^{-1}$) is highest in

a) AP b) Gujrat

c) Tamilnadu d) None of these

606. The groundnut production during 2000-01 was highest in the state of

a) Maharastra b) Gujarat

c) Andhra pradesh d) Tamil Nadu

607. The origin of rice is

a) India b) China

c) Thaland d) Bhutan

608. The *Prosopis cineraria* is commonly grows in

a) Haryana b) Rajasthan

c) M.P. d) U.P.

609. N.I. Vavilov's 'centres of origin' were based on

a) Climate b) Vegetation

c) Soil d) Geographical distribution of crops

610 The minimum support price(MSP) of Pearlmillet during 2001-02 was (Rsd^{-1})

a) 485.0 b) 610

c) 560.00 d) 690.00

611. Which one of the following weather conditions is indicated by a sudden fall in barometer reading

a) Stormy weather b) Calm weather

c) Cold and dry weather d) Hot and sunny weather

612. A radioactive substance has a half life of 4 months. Three-fourth of the substance would decay in ———————— months

a) 3 b) 4

c) 8 d) 12

613. Epiphytes are plants which depend on other plants for

a) Food b) Mechanical support

c) Shade d) Water

614. Quartizite is meamorposed from

a) Limestone b) Obsidian

c) Sandstone d) Shale

615. The earlier name of WTO was.

e) UNCTAD f) GATT

g) UNIDO h) OECD

616. The American multinational company , Monsanto has produced an insect-resistant cotton variety that is under going field trials in India. A toxin gene from which one of the following bacteria has been transferred to this transgenic cotton?

a) *Bacillus subtilis* b) *Bacillus thusigiensis*

c) *Bacillus globlii* d) *Bacillus amyloiquifanciens*

617. Which one of the following chemicals is a vitamin

a) Phosphoric acid
b) Pentothenic acid
c) Ascorbic acid
d) Succinic acid

618. The groundnut varieties developed through mutation breeding include

a) TG – 26
b) TRM – 18
c) Both of these
d) None of these

619. The 4th ministerial conference of WTO was held during November 2001 at

a) Doha
b) Seatlle
c) Geneva
d) Singapore

620. The deficiency of 'Niacin' course

a) Ricket
b) Ostemalasia
c) Pellegra
d) Night blindness

621. According to UNDP estimates, the foodgrain requirement to feed the burgeoning population in India by 2020 AD would be

a) 260 MT
b) 280 MT
c) 240 MT
d) 300 MT

622. Agricultural census is taken after every ———— year

a) 5
b) 10
c) 15
d) 3

623. The national Agricultural Technology project (NATP) is financed by

a) ADB
b) World bank
c) CIDA
d) None of these

624. The National Research center on camel is located at

a) Jorbeer
b) Modipuram
c) Avikanagar
d) Karnal

625. The descending order of NPK mobility in the soil is.

a) N>K>P
b) N>P>K
c) P>N>K
d) K>P>N

626. The term 'Bonsai' refers to

a) Japanese way of flower arrangement

b) Japanese style of training plants in miniature form

c) A Method of cut house cultivation

d) None of these

627. The heterosis over the better parent is called.

a) Relative heecrosis
b) Standard heterosis
c) Pseudoheterosis
d) None of the above

628. The "Sonalika" variety of wheat has dwarfing genes derived from

a) Rht 1 – 10
b) Stu – 1
c) Norin – 1
d) Norin – 10

629. Brinjal is considered to have had its origin in

a) China
b) Africa
c) India
d) Japan

630. The volume (in litres) of one cumec flow of water in one hour is

a) 2.6×10^6
b) 3.6×10^6
c) 4.6×10^6
d) 5.6×106

631. Which measure of central tendency is used to find the average size of shoes sold in the market.

a) A.M.
b) Median
c) G.M.
d) Mode

632. The minimum sample size for using x^2 – test should be

a) 100
b) 30
c) 50
d) 5

633. The degrees of freedom for error in RBD with 8 treatments and 4 replications will be

a) 21
b) 32
c) 16
d) 24

634. The latin square design (LSD) is suitable for companing —— treatments.

a) 3
b) 5-12
c) 2-4
d) 15

635. For testing the independence of two attributes, the test used is

a) T – test
b) F – test
c) X^2 – test
d) Z – test

636. DAP is a

a) Mixed fertilizer
b) Straight fertilizer
c) Complex fertilizer
d) Complet fertilizer

637. The growth substances which play a vital role in the division of cytoplasm are

a) Auxins
b) Gibberllins
c) Cytokinins
d) ABA

638. The source of infection of black rust of wheat in the nothemplains if India is

a) Teliosqores
b) Uredospores
c) Pycniospores
d) All of the above

639. Phoma is a .... mycorrhizae

a) Ectotrophic
b) Endotrophic
c) Either a or b
d) Both a and b

640. The crop growth rate (CGR) can be expressed as

a) NAR × LAI
b) NAR × LAR
c) LAI × duration
d) None of these

641. The most suitable herbicide for replacing blind hoeing in sugarcane is

a) 2,4 – D
b) Atrazine
c) Paraquat
Alachlord)

642. Burgundy mixture is prepared by mixing

a) $CuSO_4$ + Quicklime $Ca(OH)_2$ + Water
b) $CuSO_4$ + Na_2CO_3 + Water
c) CuSO4 + Na2CO3 + Ca(OH)2
d) None of these

643. Karnal bunt of wheat is caused by fungus

a) Ustilago nuda
b) Neovossia indica
c) Puccinia graminis
d) None of the above

644. The gram wilt can be controlled by.
 a) Late sowing b) Deep ploughing
 c) Use of resistant varieties d) All of the above

645. Stenosis / Small leaf diseare of cotton is caused by
 a) Fungi b) Bacteria
 c) Virus d) Physiological disorder

646. Bunga (*Aeginitia indica*) is root parasite on.
 a) Sugarcane b) Sorghum
 c) Mustard d) Tobacco

647. Which one of the following is a herbicide?
 a) Carbaryl b) Endoswfan
 c) Malathion d) None of these

648. The sorghum midge is
 a) Contarinia sorghicola *b) Chilr partellus*
 c) Cole mania d) None of these

649. The paddy mealy bug is
 a) Mythimna b) Ripersia oryzae
 c) Dicladispa armigera d) Tryporyza incertulas

650. Rhizome is a specialized ....
 a) Root b) Stem
 c) Either a or b d) None

651. The most destructive disease of sugarcane is
 a) Red rot b) Wilt
 c) Smut d) Albino

652. The center of origin of wheat is
 a) South western Asia b) South-Eastern Asia
 c) India d) China

653. Chickpea pod borer can be controlled by the spray of
 a) Monocrotophos @1ml/litre of water
 b) Aldrin (5%) @15 kg/ha
 c) Seed treatment by captan (0.25%) @4g/kg of seed
 d) BHC @10 kg/ha

654. The recommended optimum seed rate for pigeon pea is

a) 12-15 kg/ha
b) 20-25 kg/ha
c) 8-10 kg/ha
d) 18-20 kg/ha

655. Indian society of Agronomy was established in the year

a) 1935
b) 1945
c) 1955
d) 1965

656. Agronomy Journal is published from

a) India
b) China
c) Japan
d) America

657. NARP was implemented in January, 1949

a) To disseminate the agricultural technology
b) To strengthening the laboratory research
c) To strengthening the regional research capabilities
d) None of these

658. Chickpea and mustard yielded 10 &6 q/ha in intercropping system and 12.5 and10 q/ha in pure cropping, respectively. The LER value of above system will be

a) 0.80
b) 1.00
c) 1.40
d) 1.80

659. The dry weight of *Phalaris minor* in control (unweeded) and treated plot is 600 and 200g/m^2, respectively. What would be the weed control efficiency in percentage?

a) 150
b) 33.33
c) 66.67
d) 75

660 What would be CPE value, if 6 cm irrigation is applied at 1.2 IW/CPE ratio

a) 5 mm
b) 50 mm
c) 60 mm
d) None of these

661. International Rice Research Institute, Manila, Philippines was established in the year

a) 1950
b) 1960
c) 1970
d) 1980

662. Who was the first president of ICAR?
   a) Dr. BP Pal b) Dr. M S Swaminathan
   c) M D Habibullah d) None of these

663. Central plant protection Institute is located at
   a) N. Delhi b) Bangalore
   c) Hyderabad d) Calcutta

664. Which one is not related with sugarcane?
   a) IISR b) SBI
   c) NSI d) CSIR

665. Which is one not included in research programme of ICRISAT
   a) Chickpea b) Pea
   c) Pigeon pea d) Sorghum

666. 13. Central Institute of Subtropical Horticulture is located at
   a) Lucknow b) Bangalore
   c) Madras d) Jhansi

667. Recommended ratio for wheat +Mustard intercropping is
   a) 4:1 b) 6:1
   c) 9:1 d) 12:1

668. In synergetic yield of both crops is
   a) Higher than pure crop b) Lower than pure crop
   c) Equal to pure crop d) None of these

669. A truly scientific approach for farming was started in 1840 by
   a) Julius Von Liebig b) Sir John Bennet Lawes
   c) Sir Humphry Davy d) None of these

670. The normal lapse ate per km is
   a) 4.5 °C b) 6.5°C
   c) 8.5°C d) 10.5°C

671. The instrument used to measure total incoming radiation
   a) Thermometer b) Barometer
   c) Pyranometer d) Hygrometer

672. The concept of plant ideotype was given by
a) C M Donald
b) Dr.MS Swaminathan
c) Dr. BP Pal
d) None of these

673. Agro-climatic regional planning in India was initiated by the planning commission in the year (during 7th Five year plan 1985-90)
a) 1978
b) 1988
c) 1982
d) 1991

674. Number of Agro-ecological zones in India is
a) 16
b) 21
c) 27
d) 15

675. First agricultural scientist who became secretary to GOI
a) MS Randhawa
b) Dr.MS Swaminathan
c) Dr. BP Pal
d) O P Gautam

676. In India about .... land is covered with forest
a) 13%
b) 15%
c) 21%
d) 18%

677. First synthetic organic chemical for weed control was introduced in the year (DNOC)
a) 1918
b) 1932
c) 1926
d) 1942

678. If the fertility of a field is in scattered patches which design is suitable?
a) CRD
b) LSD
c) RBD
d) None of these

679. During the last 3 decades the estimates postulates that the CO_2 content of the atmosphere has increased from 280 ppm to
a) 250
b) 300
c) 350
d) 400

680. The minimum support price of wheat during 2002-2003 in terms of Rs/q
a) 610
b) 620
c) 640
d) 360

681. PRH-10 Basmati rice hybrid has been developed at
a) PAU, Ludhiana
b) GBPAUT, Pantnagar
c) TNAU, Coimbatore
d) None of these

682. Permisssible biuret limit in most urea fertilizers is

a) 5% b) 4%

c) 3% d) 1.5%

683. CEC of humus varies from

a) 200-300 b) 200-400

c) 200-500 d) None of these

684. Chemical which regulates opening and closing of stomata

a) Kaoline b) 2,4-D

c) PMA@10^{-4}M conc d) Paraquat

685. Which one of following is the special scheme launched by ICAR for transfer of technology to the farmers

a) All India Coordinated Research Projects 1957

b) Operational Research Projects, 1974

c) Lab to Land Programme, 1979

d) National Demonstration Programme, 1964

686. Which group of microbes increases phosphorus solubility?

a) Azotobacter *b) Pseudomonas*

c) *Clostridium* d) *Rhizobium*

687. The principle of experimental design includes

a) Replication b) Randomization

c) Local control d) All of these

688. The first inter genetic cross between bread wheat and rye (Triticale) was made by

a) Karpechenko,(1928) b) Rimpu (1890) at Sweden

c) Mutzing (1951) d) Andrew Knight (1935)

689. White bud of Maize is caused due to the deficiency of

a) Al b) Cu

c) Zn d) Mn

690. Three varieties of wheat with 4 N levels are replicated thrice in an experiment under randomized block design, the error degree of freedom will be

a) 12 b) 18

c) 22 d) 35

691. Dry matter production of maize increased from 500g/m^2 at 30 days of growth to 750g/m^2 at 45 days of growth. Find out the crop rate

a) 25g/day/m^2 b) 27.77g/day/m^2

c) 8.33g/day/m^2 d) 16.66g/day/m^2

692. A weed plant especially undesirable, troublesome and difficult to control is known as

a) Absolute weed b) Facultative weed

c) Noxious weed d) Troublesome weed

693. Which among the following is a long duration photosensitive variety of rice

a) IR-8 b) Jaya

c) IR-36 d) Savitri

694. The nitrogen uptake of rice increased from 50 kg N/ha under control (no nitrogen) to 70 kg N/ha with application of 40 kg N/ha. Find out the apparent N recovery

a) 50.00% b) 57.14%

c) 80.00% d) 60.00%

695. Total N content of a soil is 0.05%. Find out the quantity of N in one hectare field, considering 15 cm soil depth and 1.5g/cc bulk density of soil

a) 1125 kg/ha b) 112.5 kg/ha

c) 11250 kg/ha d) 1250 kg/ha

696. Which among the following is a stable isotope?

a) N^{15} b) C^{14}

c) P^{32} d) Zn^{65}

697. The chemical formula of urea is

a) $CO(NH)_2$ b) $CO(NH_2)_2$

c) CO_2NH_2 d) $(CO)_2NH_2$

698. Siemen per metre is a unit for measurement of

a) Electrical resistance b) Electrical conductivity

c) Hydraulic conductivity d) Exchangeable sodium

699. Pressmud is byproduct of

a) Sugar industry b) Paper industry

c) Steel industry d) Sulphuric acid

700. A soil test report revealed pH=8.0, Electrical conductivity=12ds/m and Sodium absorption ratio=3.0. What is the kind of soil

a) Normal soil
b) Saline soil
c) Saline – sodic soil
d) Sodic soil

701. Who in 1865 coined the original Greek term 'Ecology' by combining two Greek word *oikas* meaning house and *logos* meaning study of

a) Hanns Reiter
b) Ernst Haeckel
c) Odum
d) None of these

702. Who for the first time defined ecology in 1866

a) H. Reiter
b) Ernst Haeckel
c) Odum
d) Krebs

703. The decrease in temperature with altitude is known as

a) Temperature inversion
b) Lapse rate
c) Thermal anomaly
d) None of these

704. The part of the earth in which life exists is.

a) Hydrosphere
b) Lithosphere
c) Biosphere
d) Atmosphere

705. Which one of the following dam has been built on the river Nice

a) Panchet
b) Tehri
c) Aswan
d) All of the above

706. The term 'Ecosystem' was proposed by the British Ecologist in 1935

a) Karl mobius
b) A.G. Tansley
c) S.A. Forbes
d) None of these

707. The tendency of natural ecosystems to resist change and to remain in a state of equilibrium is known as

a) Homeostatis
b) Biotic potential
c) Ecosystem
d) None of these

708. The 'law of tolerance' was proposed by

a) V.E. Shelford
b) J.V. Liebig
c) C.A. Black
d) All of these

709. The visible light to human eye is within the wavelength of

a) 3900 – 7600 A° b) 390 – 760 nm

c) 0.39 – 0.76 μ d) All of the above

710 Which of the following lights are most useful for photosynthesis

a) Green and Yellow b) Blue and Red

c) Violet and infrared d) Green and Orange

711. The high concentration of nitrate in drinking water has been found to cause ——— in infants

a) Methaemoghibinaemia b) Eutrophication

c) Osteomalacia d) None of these

712. The world health organisations recommended safe limit for potable water is only – mg/l of nitrate

a) 25 b) 35

c) 45 d) 55

713. The headquarters of world wide fund for nature (WWF) is situated at

a) Glands, Suitzerland b) Newyork, USA

c) Paris, France d) Naples, Italy

714. the project tiger was launched in

a) 1971 b) 1972

c) 1973 d) 1974

715. When was Dr. Norman E. Borlaug was awarded the Noble peace prize for his work in agriculture

a) 1970 b) 1971

c) 1972 d) 1973

716. The soil erosion rate (tonne / ha) in India is about

a) 10.35 b) 16.35

c) 22.35 d) 28.35

717. Osmotic Tensiometers developed in Australia can measure soil tension upto ——bars

a) 0.85 b) 8.5

c) 20 d) 33

718. The widely used method of soil loss estirnation, universal soil loss Equation (USLE) given by A = RKLSCP, was proposed by ———— in 1960

a) Wischmeier and Smith
b) Middleton
c) Chepil
d) Schofield

719. Who proposed the hydrometer method for mechanical analysis of soil separates

a) Mick
b) Bouyoucos
c) Atterberg
d) Brounauer

720. The pF range of moisture found optimal for tillage operation in medium – textured soils is

a) 1.4 – 2.8
b) 2.8 – 4.4
c) 4.4 – 6.6
d) 6.6 – 8.8

721. A group of soils which have developed from the some parent material in same climate but under different topographical condition is called

a) Soil catena
b) Cat clays
c) Clay micelle
d) Podzol soils

722. The plants which can tolecate extreme salinity are classed as

a) Harophytes
b) Oligohalophytes
c) Euchalophytes
d) Glycophytes

723. "The resistance offered by a liquid medium against the fall particles down the same liquid medium, varies with the radius of the particles" is an expression of

a) Fick's law
b) Stoke's law
c) Darcy's law
d) None of these

724. Calculate the weight of one cubic foot of a dry soil , if its bulk density is 1.2 Mg/m^3

a) 62.44 lbs
b) 74.93 lbs
c) 28.3 lbs
d) 44.1 lbs

725. In a water molecule, two hydrogens are connected with one oxygen molecule at an angle of

a) 105^0
b) 110^0
c) 100^0
d) 115^0

726. Water molecule is ———— in nature

a) Polar b) Dipolar
c) Multipolar d) None of these

727. The energy involved in the retention of water by the soil is mainly

a) Kinetic energy b) Potential energy
c) Both of the above d) None of these

728. What will be percentage base saturation, if CEC of the soil is 10 cmol / kg and it has absorbed 8 milliequivalents of basic cations

a) 60 b) 80
c) 100 d) 40

729. The high solute content in rooting medium create water stress by ——— in osmotic potential

a) Decrease b) Increase
c) Both of these d) None of these

730. The symbiotic association between algae and fungi is called

a) Lichens b) Mycorrhiza
c) Commensalism d) None of these

731. The C : N ratio of organic matter usually ranges from

a) 8 : 1 to 15 : 1 b) 40 : 1 to 60 : 1
c) 60 :1 to 100 : 1 d) 4 : 1 to 9 : 1

732. The bacteria used for symbiotic nitrogen fixation in Pigeonpea is

a) *Cowpea Miscellany* b) *leguminosarrum*
c) *R. phaseoli* d) R. lupini

733. Which one of the following first suggested the genetic system of soil classification

a) C.F. Marbut b) V.V. Dokuchaev
c) Atterberg d) None of these

734. The reaction 2 Fe_2O_3 ! 4 FeO + O_2 is an wxample of

a) Oxidation b) Reduction
c) Hydration d) Hydrolysis

735. The fossil fuels are excavated from

a) Igneous rocks b) Sedimetary rocks
c) Metamorphic rocks d) All of these

736. The igneous rocks containing < 45 % silica are classified as:

a) Basic rocks
b) Acidic rocks
c) Sub-basic rocks
d) Ultra-basic rocks

737. The CO_2 content of atmospheric air on weight basis (%) is about

a) 0.03
b) 0.04
c) 0.05
d) 0.06

738. Who said that 'Humus' is the principle of vegetation

a) Wallerius
b) Glauber
c) Woodward
d) Ignehousz

739. The gases formed by the anaerobic decomposition of organic matter is

a) H_2
b) CH_4
c) Both of these
d) None of these

740. Which one of the following is key element for the synthesis of carbohydrates

a) C
b) H
c) O
d) N

741. Which one of the following is the key element for the synthesis of protein

a) C
b) H
c) P
d) N

742. At Rothamsted, in England, J.B. Lawes and J.H. Gilbert (1842) produced superphosphate by chemical treatment of crushed bones with

a) HCl
b) H_2SO_4
c) HNO_3
d) H_2CO_3

743. Who discovered the antibiotics streptomycin in 1944

a) Alexander Flemming
b) S.N. Winogradsky
c) Waksman
d) None of these

744. Who developed the concept of phyllosphere

a) Ruinen
b) Van Niel
c) Rossi
d) None of these

745. Who demonstrated that an adequate supply of molybdenum was essential for accelerating nitrogen fixation by nodulating legumes

a) Kudo
b) Bortels
c) Virtanen *et al*
d) None of these

746. Who for the first time reported that root nodules of legumes could be effective only in the presence of a red pigment in them

a) Kudo b) Bortels

c) Gibson d) Vincent

747. The term rhizosphere was introduced in 1904 by the

a) Hiltner b) Vincent

c) N.S. Subba Rao d) None of these

748. The molecular weight of laghaemoglobin is of the order of

a) 16,000 – 17,000 b) 66,000 – 67,000

c) 40,000 – 41,000 d) None of these

749. The spraying of plants with urea ——— nodulation and N_2 fixation

a) Prevents b) Enhances

c) Doesn't affect d) None of these

750. The spraying of plants with sucrose ——— nodulation and N_2 fixation

a) Prevents b) Enhances

c) Doesn't effect d) None of these

751. Which one of the following is not a Sulphur oxidizing bacteria

a) *Thiobacillus thiooxidans* b) I. Thioparus

c) *Desulfovibrio disulfuricans* d) None of these

752. The term 'Mycorrhiza' literally means

a) Fungus with association with roots b) Plant roots

c) Leaf surface d) None of these

753. The 'Biosuper' is a form of ——— fertilizer

a) Organic b) Inorganic

c) Both of the above d) None of these

754. The highest limit of active acidity is at pH

a) 14.0 b) 7.0

c) 0.0 d) 1.0

755. A solution with pH 5.0 is ——— times as acdic as that having pH 4.0

a) 10 b) 1/10

c) 1/100 d) 100

756. The highest percentage porosity is of
   a) Alluvial soil   b) Black soil
   c) Lateritc soil   d) Red soil

757. The soils containing ——— percentage of clay are undesirable for irrigated farming
   a) < 50   b) > 50
   c) < 5   d) > 5

758. The irrigation water is classified as medium SAR water if the SAR is
   a) <10   b) 10-18
   c) 18-26   d) >26

759. The electrochemical theory of nutrient uptake was advocated by
   a) Lundegarh   b) Milne
   c) Parker   d) None of these

760 The degree of saturation of the soil exchange complex with sodium is called
   a) SAR   b) ESP
   c) PAR   d) EC

761. Who stated that "plants with high nutrient requirement also had a high water requirement"
   a) Theophrastus   b) Jethro Tull
   c) Pietro de Crescenzi   d) Liebig

762. The prefix WH in wheat varieties stands for
   a) Wheat hybrid   b) Wheat hoshiyarpur
   c) Wheat hisar   d) Wheat

763. Who is known as "Mr. Rice" in international circle
   a) G.S. Khush   b) S.K. De Dutta
   c) M.V. Rao   d) None of these

764. The rate of flow of liquid or flux through a posous medium is proportional to the hydraulic gradient in the direction of flow of the liquid is an expression of
   a) Darcy's law   b) O.W. Wilcox's law
   c) Graham's law   d) None of these

765. The mustard can be protected against frost by applying
a) Irrigation b) Herbicide
c) Fertilizer d) None of these

766. Urea is ———— acidic as compared to ammonium sulphate
a) More b) Less
c) Equal d) None of these

767. The main structural elements of plant tissues include
a) P, Ca, Mg b) N, P, S
c) C, H, O d) Fe, Mn, Zn

768. Which one of the following nutrient is not absorbed by plants as salts
a) Nitrogen b) Phosphorus
c) Boron d) Potassium

769. Which is the major plant nutrient lost by leaching
a) N b) P
c) K d) All of these

770. The chemical formula of dicalcium phosphate is
a) $Ca\ (H_2PO_4)_2$ b) $Ca\ HPO_4$
c) $Ca_3(PO4)_2$ d) None of these

771. The percent nitrogen content of calurea is
a) Flowering b) Milking
c) Maturity d) Gerunination

772. The percent nitrogen content of calurea is
a) 46.0 b) 34.0
c) 38.0 d) 31.8

773. The gypsum and monocalcium phosphate are present in SSP in the ratio of
a) 3 : 2 b) 2 : 3
c) 1 : 2 d) 2 : 1

774. Under which conditions, rock phosphate is also used as a fertilizer straightway
a) Finely grounded fertilizer
b) Perennial crops
c) Strongly acidic soils and high humus content
d) All of these

775. The crops which utilize most efficiently phosphorus from rock phosphate include

a) Turuip and sweet clover b) Palak an spinach

c) *Brassica* sp d) All of these

776. The organic radical (chelating agent) in chelates is known as the — which is negatively charged

a) Ligand b) Micelle

c) Catena d) None of these

777. The protein and oil content (%) of winged – beaus is —— respectively

a) 33, 18 b) 40, 20

c) 25, 10 d) 23, 13

778. The oil and protent content (%) of Jojoba ranges from —— and ——— respectively

a) 45 – 55; 30 – 37 b) 35 – 45 ; 30 – 37

c) 30 – 37 ; 45 – 55 d) None of these

779. The Asia's largest grain market is located at

a) Khanna, Ludhiana b) Kekri, Ajmer

c) Hapur, Mujaffarnagar d) None of these

780. Which one of the following variety is of triticale developed by IARI

a) Trishulata b) Vasun

c) Madhuri d) DT 46

781. Which is the first variety of mustard in the world developed by using biotechnology

a) ICMH 356 b) Pusa 322

c) Pusa Jaikisan d) Uday

782. Which one of the following is the most economical and promising herbicide for controlling the weeds of transplanted rice

a) Anilophos b) Benthiocarb

c) Butachlor d) Isoproturon

783. Wheat is usually hawested in the month of :

a) July b) September

c) December d) April

784. Green revolution in India was set in by

a) Maize hybrids
b) Dwarf wheat
c) HYV's of rice
d) All of these

785. The largest total production of wheat is in the state of

a) Punjab
b) Haryana
c) UP
d) Rajasthan

786. The 'ratio law' was enunciated by

a) Beckett
b) Schofield
c) Russell
d) Barber

787. The high yield wheat variety which is specifically developed for growing on the manganese deficient soils of the northern plains is

a) HD 2428
b) HD 2329
c) Raj 3777
d) All of these

788. UO 285 and kirtiman are the varieties of

a) Opium
b) Asgand
c) Safed musli
d) None of these

789. 2,4 – dichlorophenoxy acetic acid is a chemical name of

a) 2,4 – DB
b) MCPA
c) 2,4 – D
d) MCPB

790. 2,2 – dichloropropionic acid is a chemical name of

a) Biuron
b) Dalapon
c) Propanil
Paraquat d)

791. The scientific name of medicinal plant 'Guggul' is

a) *Rauwolfia serpentina*
b) Commifora mukal
c) *Atlopa belladonna*
d) Papaver somniferum

792. The rotational intensity of groundnut – Pigeonpea – sugarcane rotation is

a) 150 %
b) 100 %
c) 300 %
d) 166 %

793. The triple gene dwarf varieties of wheat include

a) UP 301 and UP 319
b) Heera and Moti
c) K 816 and UP 310
d) All of these

794. The chief source of energy in the sun is

a) Fission reaction b) Fusion reaction

c) Chemical reaction d) Burning reaction

795. The Green house effect (GHG) ——— the atmospheric temperature

a) Increases b) Decreases

c) Doesn't affect d) Fluctuates

796. The acreage (million hectares) under sugarcane cultivation in India is approximately

a) 3.0 b) 4.0

c) 5.0 d) 6.0

797. The largest number of 'Sugar mills' are in the state of

a) U.P. b) M.P.

c) Maharastra d) Tamil Nadu

798. The 'Borlaug award' instituted by Corromandal Fertilizers Limited and given in the field of agriculture was started in the year

a) 1963 b) 1973

c) 1983 d) 1993

799. The essentiality of which nutrient has been demonstrated for symbiotic N – fixation in soybean, lucerne and subterranean clover

a) Na b) Fe

c) Co d) Mg

800. The 88th session of 'Indian Science Congress' held at IARI ,New Delhi during 2001 was presided over by

a) R.S. Paroda b) Panjab Singh

c) A.S Faroda d) M.S. Swaminathan

801. The 'Sampurna Gramin Rozger Yojana ' (SGRY) was launched on

a) 15 August, 2001 b) 25 September, 2001

c) 26 January, 2002 d) 1 April, 2002

802. Which genera of VAM have been observed in the roots of grasses, cueals and legumes?

a) *Endogone* spp. b) *Exogone* spp.

c) *Monotropa* spp. d) None of these

803. The negative logarithm of electron activity is referred to as
a) pH b) pF
c) pE d) None of these

804. Chroma indicates
a) Brightness of soil colour b) Purity of soil colour
c) Dominant spectrum d) All of these

805. Dr. Hergovind Kharana was awarded the Nobel prize for in vitro synthesis of gene in the year
a) 1958 b) 1968
c) 1979 d) 1983

806. A disease which occurs widely but periodically is termed as
a) Endemic b) Epophytotic
c) Sporadic d) Pandemic

807. The flag smut of wheat is caused by
a) Urocystis tritici b) Neovossia horrida
c) Neovossia indica d) None of these

808. One nanometer (nm) is equal to
a) 0.1 μ b) 0.01 μ
c) 0.001 μ d) 0.0001 μ

809. According to ISTA, the minimum number of seeds required for germination test are
a) 100 b) 200
c) 300 d) 400

810 Find out the area sprayed by a full tank, if the volume of the tank is 300 litres and rate of spray is 150 litres ha^{-1}
a) 0.5 hectare b) 2.0 hectare
c) 1.0 hectare d) None of these

811. The weed which is used against snake bite is
a) *Leucas sapera* b) *Argemone maxicana*
c) *Cypesus rotundus* d) *Saccharum* spp.

812. What will be the real value of seed if the seed sample has 80 per cent purity and 70 per cent germination
a) 78 % b) 56 %
c) 80 % d) None of these

813. The khaira disease of paddy could be controlled with the basal application of

a) 25 Kg $ZnSO_4$ ha^{-1}
b) 5 Kg $ZnSO_4$ + 2.5 Kg lime ha^{-1}
c) Both of these
d) None of these

814. Find out the seed rate of wheat (Kgha-1) if spacing is 20 cm × 3 cm , germination % is 95, purity percentage 90 and test weight is 46 g

a) 70
b) 80
c) 90
d) 100

815. The female fertive parent used for hybrid seed production is

a) 'B' line
b) Maintainer line
c) 'A' line
d) Both a and b

816. The 'Pahala blight of sugarcane' is caused by the deficiency of

a) Zn
b) Mn
c) Cu
d) Fe

817. Find out the cost per Kg of ammonium sulphate containing 20 % nitrogen and the cost of

a) Rs. 2.0
b) Rs. 0.5
c) Rs. 5.0
d) None of these

818. An aggeessivity value of—————— indicates that both the species in mixed / intercropping systems are equally competitive

a) 0
b) -
c) +
d) None of these

819. The relative area of sole crop required to produce the yield obtained from mixed / intercrops is termed as

a) RCC
b) MCI
c) LER
d) Aggressivity

820. The practice of grousing only one crop in a year is called

a) Monocropping
b) Monoculture
c) Intercropping
d) Multiple cropping

821. The department of agricultural rerearch and education (DARE) was created in the ministry of Agriculture in ——

a) July, 1973
b) December, 1973
c) July, 1879
d) December, 1979

822. Which is the rational stage of production in classical production function
a) First stage b) Second stage
c) Third stage d) Both b and c

823. Sucrose is a
a) Reducing sugar b) Non-reducing sugar
c) Sugar with both property d) None

824. The operational size of holding in India is ——— hectares
a) 1.55 b) 1.65
c) 1.75 d) 1.45

825. The second international crop science congress was held in New Delhi during
a) April, 1996 b) November, 1996
c) April, 2000 d) November, 2000

826. The National Institute of Animal Nutrition and Physiology was located at
a) Bangalore b) Hyderabad
c) Mumbai d) Kolkata

827. The Fakhruddin Ali Ahmed award was given for outstanding agricultural research in
a) *Desert* areas b) *Tribal* areas
c) *Hilly* areas d) None of these

828. The Institute village Linkage programme (IVLP) was launched by
a) IARI b) ICRISAT
c) ICAR d) ICARDA

829. Who is the creator of super rice
a) M.S. Swaminathan b) G.S. Khush
c) S.K. De Dutta d) None of these

830. Prof. P.V. Sukhatme was associated with
a) Agril. Statistics b) Soil science
c) Agronomy d) None of these

831. The headquarters of Marine products Export Development Authority is located at
a) Mumbai b) Cochin
c) Chennai d) Bhubaneshwar

832. The ICAR was set up on the recommendation of
   a) Royal commission on Agriculture
   b) National commission on Agriculture
   c) Both of these
   d) None of these

833. Who is the president of the ICAR
   a) A.B. Vajpayee b) Ajit Singh
   c) H.D.N. Yadav d) Panjab singh

834. The National Agricultural Research Project (NARP) was launched in
   a) January, 1979 b) December, 1979
   c) January, 1973 d) December, 1973

835. DNA synthesis occurs during
   a) Metaphase b) Interphase
   c) Prophase d) GR phase

836. Among the following which is not a programming language
   a) BASIC b) FORTRAN
   c) IBM d) COBOL

837. The most important aspect of extension programme planning is
   a) Participation of local people in the programme planning
   b) Provision of adequate finance
   c) Availability of marketing facilities for the produce
   d) Involvement of officials in the programme planning

838. Which one of the following is not electromagnetic radiation
   a) Gamma rays b) Alpha rays
   c) X – rays Radio – wavesd)

839. The ICAR celebrated its Golden Jubilee in
   a) 1979 b) 1954
   c) 1939 d) 1949

840. The CIPHET is located at
   a) Lucknow b) Ludhiana
   c) New Delhi d) Bhopal

841. The IARI was shifted from pusa, Bihar to New Delhi in the year
a) 1905 b) 1934
c) 1936 d) 1929

842. Which fruit crop has the highest productivity (t/ha)
a) Banana b) Papaya
c) Crape d) Mango

843. The journal 'Indian Horticulture' is published by
a) ICAR b) IARI
c) Hort. Soc. Of India d) CSIR

844. The quincunx system accommodates ——— times more plants than square system of planting
a) About 2 b) About 1.5
c) About 3 d) None of these

845. The insecticides of plant origin includes
a) Pyrethrum and Andsim b) Endrin and Rotinon
c) Rotinon and pyrethrum d) Endrin and Aldrin

846. Pore space in sandy soil
a) 5-10% b) 10-20%
c) 20-25% d) 20-30%

847. Which state has the highest net sawn area in the country
a) Maharastra b) UP
c) Rajasthan d) MP

848. Which one of the following is not an organomercurial fungicide
a) Agrosan GN b) Ceresan
c) Parrugen d) Demosan

849. The diseased grains partly converted into black sooty powder which smells like rotten fish is a characteristic symptom of wheat diseare
a) Karnal bunt b) Loose smut
c) Stripe rust d) All of these

850. A systematic infection 'Kresek' is associated with which disease of rice
a) Ufra b) Blast
c) Brown spot d) BLB

851. The basic structural and functional unit of life is
a) Cell
b) Tissue
c) Organ
d) None of these

852. The functions of cell wall include
a) Mechanical support
b) Absorption and transport of water and minerals
c) Disease resistance
d) All of these

853. The most abundant chemical present in secondary cell walls is
a) Cellulose
b) Lignin
c) Hemi-cellulose
d) Suberin

854. The forms of passive transport include
a) Diffusion
b) Ion exchange
c) Mass flow
d) All of the above

855. The contents of the vacuoles are collectively referred to as the
a) Cell sap
b) Cell contents
c) Cytoplasm
d) Protoplasm

856. The primary function of spherosomes in the cell appears to be the
a) Metabolism of the glycolate
b) Storage and transport of lipods
c) Control of protein synthesis
d) All of the above

857. Which of the following cell structures are included in microbodies
a) Glyoxysomes
b) Peroxisomes
c) Spherosomes
d) All of the above

858. The reaction of krebscycle occur in
a) Cholroplart
b) Ribosomes
c) Mitochondria
d) None of the above

859. Glyoxysomes are primarily found in tissues of
a) Oil – bearing seeds
b) Protein- bearing seeds
c) Both of these
d) None of the above

860 7. Which one of the following is the strongest bond
a) Hydrogen bond
b) Van der waals force
c) Covalent bond
d) Ionic bond

861. The pure water is said to be ———— to the sucrose solution
a) Hypertonic
b) Hypotonic
c) Isotonic
d) None of the above

862. The area of the root where most water absorption takes place is :
a) Root hair zone
b) Cortex
c) Epidermis
d) Root cap

863. The principal pathway by which water is translocated in the angiosperms is
a) Tracheids
b) Vesrel system
c) Xylem parenchyma
d) Fiber

864. The terms apoplant and symplart were originally introduced by
a) Munch
b) Stocking
c) Dixon
d) Kramer

865. If water moves in, the cells become
a) Flaceid
b) Turgid
c) Both of these
d) None of the above

866. The part of a plant cell or tissue that allows free diffurion to take place is referred to as
a) Outer space
b) Inner space
c) Apparent free space
d) Bound space

867. 13. The carrier concept of ion transport was formulated by
a) Epstein
b) Van den Honest
c) Hagen
d) Undegardh and Burstrom

868. Which mineral elements is a constituent of ATP and other high – energy compounds
a) P
b) Ca
c) Zn
d) Fe

869. 13. The prokaryotic cells are present in
a) Cyanobacteria
b) Mycoplasma
c) Bacteria
d) All of the above

870. The casparian strips are present in
a) Endodermis
b) Epidermis
c) Cortex
d) Xylem

871. The most commonly grown crop plants are

a) Halophytes b) Glycophytes

c) Petrophytes d) Hydrophytes

872. The general formula of carbohydrates is

a) $(CH_2O)_n$ b) $(C_2HO)_n$

c) $(CHO_2)_n$ d) (C2H2O)n

873. Which of the following is / are pent sugar (s)

a) Ribose and Deoxyribose b) Glucose and Fructose

c) Dextrose and sucrose d) All of the above

874. A much smaller, organic non – protein portion of an enzyme

a) Coenzyme b) Prosthetic group

c) Isozyme d) Holoenzyme

875. Which is not an excellent example of determinate plant structure

a) Flowers b) Fruits

c) Leaves d) Roots

876. Which of the following is not a bienuial

a) *Beta vulgaris* b) *Daucus carota*

c) *Hyoscyamns niger* d) *Agave amesicana*

877. 13. Which of the following is not a bienuial

a) IAA b) IBA

c) NAA d) 2,4-D

878. 13. The Gibberellins were first discovered in

a) India b) Japan

c) USA d) Canada

879. The light intensity at which photosynthesis and respiration are equal is called

a) CO_2 compensation point b) Light compensation point

c) Light saturation point d) All of the above

880. Which one of the following is a secondary plant nutrient

a) N b) Ca

c) Zn d) Mn

881. The physiological disorder 'water core' of turnip is caused by the deficiency of

a) B b) Ca

c) Zn d) Mn

882. 2. Which is a non – reducing sugar

a) Glucose b) Fructose

c) Sucrose d) All of the above

883. The maximum photosynthetic rate of C_4 plant is types under natural conditions is of the order —— ì mol CO_2 m^{-2} S^{-1}

a) 0.6-2.4 b) 10-20

c) 20-40 d) None of the above

884. The CO_2 compensation point (ppm) of C_4 plants is

a) 0-10 b) 30-70

c) 0-5 d) 100-200

885. The ratio of leaf area of a crop to the ground area upon which that crop grows is referred to as

a) LAR b) LAI

c) NAR d) RGR

886. The process by which N_2 is reduced to NH_4^+ is called :

a) Nitrogen fixation b) Nitrogen assinerlation

c) Ammonification d) None of the above

887. The *Rhizobia* are ———————— bacteria

a) Anaerobic b) Aerobic

c) Facultative anerobes d) None of the above

888. An upward bending of a leaf or any organ is called

a) Hyponarty b) Epinasty

c) Pulvinus d) Totipotency

889. "The power of movement in plants" is written by

a) Charles Dauvia and Francis Darwin

b) Lamarck

c) Salisbury and Ross

d) None of the above

890. The ability of one to a few cells to regenerate an entire plant is called

a) Juvenility
b) Totipotency
c) Maturity
d) None of the above

891. The low temperature promotion of germination is referred to as

a) Vervalization
b) Stratification
c) Scarification
d) Photoperiodism

892. Who coined the term florigen

a) M. Chailakhyan
b) Sachs and Hackett
c) Bunning
d) None of the above

893. The effects of growth conditions carried over two or more generations is called

a) conditioning effects
b) Carry – over effects
c) Homeostasis
d) Triggered effect

894. Any change in environmental conditions that might reduce or adversely change a plant's growth or development is referred to as

a) Biological stress
b) Biological strain
c) Physical stress
d) Physical strain

895. The example of strain includes

a) Reduced light levels
b) Frost
c) Reduced photosynthesis
d) All of the above

896. "The rate of a chemical reaction is doubled by increasing the temperature 10°C" is an expression of

a) Beer lambert's law
b) Van't Hoff's law
c) Henry's law
d) Ohm's law

897. The study of the relationships between crop plants and their environment is known as

a) Crop ecology
b) Ecological crop geography
c) Ecology
d) Agrology

898. Rice may be considered as a

a) Facultative hydrophyte
b) Obligate hydrophyte
c) Xerophyte
d) Mesophyte

899. Photosynthesis is an —————————— process

a) Oxidation b) Reduction

c) Oxidation reduction d) None of the above

900. Which nutrient element is involved in the biosynthesis of IAA?

a) Fe b) Zn

c) Cu d) Mn

901. The united states patent office has given a patent to a Texan company for selling some varieties of ——————

a) Basmati rice b) Turmeric

c) Papaya d) Neem

902. Which of the following sectors contributes maximum in deciding the growth in income of the states of India?

a) Energy b) Transport

c) Agriculture d) Tourism

903. Which of the following is the state where the number of people living below poverty line is?

a) Bihar b) Orissa

c) Rajasthan d) UP

904. Who is the person closely associated with operation flood programme and was honored by padma vibhushan recently?

a) Dr. V. Kurien b) Dr. M.S. Swaminathan

c) Dr. Amartya Sen d) None of these

905. Which one of the following is a fat soluble vitamin?

a) Vitamin A b) Vitamin B

c) Vitamin C d) Vitamin D

906. The wound healing vitamin is

a) Vitamin E b) Vitamin D

c) Vitamin C d) Vitamin A

907. Which vitamin is known as an "Anti – sterility vitamin"?

a) Vitamin D b) Vitamin E

c) Vitamin B d) Vitamin K

908. The method of growing of vegetables out of their normal season is called

a) Vegetable forcing b) Truck gardening

c) Market gardening d) None of these

909. Which one of the following is the 'American Bollworm' of cotton?

a) *Earias insulana* b) *Helicoverpa armigera*

c) *Pectinophora gossypiella* d) *Aphis gossypii*

910 Which state is the largest producer of mango in India?

a) UP b) AP

c) Rajasthan d) Haryana

911. The powdery mildew of mango is caused by

a) *Oidium mangiferal* b) *Phoma glomerata*

c) *Lasiodiplodia theobromal* d) All of these

912. The largest producer of pigeonpea is

a) MP b) UP

c) Rajasthan d) Haryana

913. The ideal conditions for survival and breeding of white grubs are

a) Light sandy well drained soils

b) Light sandy water-logged soils

c) Heavy clay well-drained soils

d) Heavy clay water logged soils

914. The pesticides banned in India include

a) BHC and Aldrin b) MSMA and Nitrofen

c) PCP and Tetradifon d) All of these

915. The pesticides for restricted use in India include

a) DDT , Dieldrin and Methyl parathion

b) Carbaryl and Lindane

c) Monocrotophos

d) All of these

916. The National Research Center for Women in Agriculture is situated at

a) Kolkata b) Bhubaneswar

c) Mumbai d) Cochin

917. The varieties developed at the IARI include

a) CSH – 106
b) Aruna
c) Kisan
d) All of these

918. Who said that "Nature provides for everybody's need but not for everybody's greed"

a) Mahatma Gandhi
b) J.L. Nehru
c) M.S. Swaminathan
d) A.B. Vajpayee

919. The world's first tungro – virus resistant rice variety is

a) Lunishree
b) Vikramarya
c) Gayatri
d) Sneha

920. The first wheat variety to have manifested substantial resistance against the three rusts was

a) NP 829
b) NP 770
c) NP 809
d) NP 4

921. The 'Norin 10' dwarfing gene in wheat was introduced from the semi – dwarf varieties of

a) Japan
b) USA
c) Mexico
d) India

922. Which wheat variety has paved way for self sufficiency in foodgrains in sudan:

a) HD 2009
b) HD 2172
c) PBW 343
d) WL 711

923. Which Indian variety of cotton is comparable with GIZA 45, the best Egyptian

a) Hybrid 4
b) Varalaxmi
c) Suviu
d) Fateh

924. Which of the following is a hybrid variety of mango

a) Arka Aruna
b) Arka puneet
c) Arka Anmol
d) All of these

925. Which one of the following is a pubeication of ICAR

a) Indian Farming
b) Kheti
c) Phal phool
d) All of these

926. The Royal commission on Agriculture was set up in the year of

a) 1926 b) 1929
c) 1947 d) 1965

927. In which year, the ICAR became the nodal agency for co-ordinating agricultural research in the country

a) 1947 b) 1965
c) 1973 d) 1979

928. The 'world Forestry Day' is celebrated every year on

a) 21 march b) 16 September
c) 23 march d) 16 October

929. The "Dee–Gee–wu–Gen" dwarfing gene in rice was sourced from which variety

a) TN – 1 b) Jaya
c) IR – 8 d) Peta

930. The Taichung Native 1 was introduced from

a) Japan b) Mexico
c) Korea d) Taiwan

931. The number of deemed to be universities in ICAR are:

a) 4 b) 5
c) 2 d) 3

932. The NRC for DNA Fingerprinting is situated at

a) New Delhi b) Hyderabad
c) Mumbai d) Lucknow

933. The tomato is a ————————— season crop

a) Warm b) Cool
c) Warm as well as cool d) None of these

934. The Asian vegetable Research and Development center (AVRDC) is located in

a) Japan b) China
c) Taiwan d) India

935. Which one of the physiological disorder is associated with tomato

a) Puff b) Catface
c) Sunscalding d) All of these

936. The most serious virus diseare of tomato in India is
   a) Mosaic b) Leaf curl
   c) Fern leaf d) None of these
937. The mango fruits are rich source of
   a) Vitamin A b) Vitamin B
   c) Vitamin C d) Vitamin E
938. The sweetest variety of mango is
   a) Langra b) Chausa
   c) Alphanso d) Amrapali
939. Which one of the following mango variety is poryembryonic?
   a) Olour b) Goa
   c) Chandrakaran d) All of these
940. The variety norin 10 as the source of dwarfing genes in wheat is a —— —— variety
   a) Japanese b) Mexican
   c) Indian d) . American
941. The gene(s) responsible for height reduction in semi dwarf wheat varieties
   a) Rht 1 b) Rht 2
   c) Both of these d) Norin 10
942. Who postulated the *Law of Homologous Series in variation*?
   a) N.I. Vavilov b) Charles Darwin
   c) G.H. Shull d) Johansen
943. The Destructive Insects and pests (DIP) Act was passed in
   a) 1914 b) 1966
   c) 1934 d) 1947
944. The headquarters of the Botanical Survey of India is at
   a) Dehradun b) Kolkota
   c) New Delhi d) Mumbai
945. The tomato variety derived from a cross between Meeruty and Sioux, an introduction from
   a) Pusa sheetal b) Pusa Gaurav
   c) Pusa Ruby d) Pusa 120

946. The potato tuber moth was introduced in 1900 from

a) Italy
b) Srilanka
c) Europe
d) USA

947. The characters Produced by polygenes are referred to as characters

a) Quantitative
b) Qualitative
c) Both of these
d) None of these

948. The application of statistical procedures to the study of biological problems is referred to as

a) Biometry
b) Geometry
c) Plantometry
d) None of these

949. Acid is a

a) Proton donor
b) Proton acceptor
c) Both a and b
d) None of these

950. The total number of meteorological sub-divisions in the country is

a) 25
b) 30
c) 35
d) 40

951. Scientific name of broad bean

a) *Vigna aconitifolia*
b) *Lablab purpureaus*
c) *Vigna umbellata*
d) *Vicia faba*

952. N.G. Dastane is related to

a) Irrigation Agronomy
b) cropping system
c) Irrigation Engineering
d) IPM

953. King of spices

a) Cardamom
b) Black pepper
c) Clove
d) Turmeric

954. Equivalent acidity of Ammonium chloride

a) 80
b) 93
c) 110
d) 128

955. Concept of pE was given by

a) Silen
b) Sorensen
c) Francis Moore
d) None of these

956. First Agriculture University in India

a) PAU
b) IGKV
c) GBPUAT
d) OUAT

957. The largest producer and consumer of tea in the world is

a) Brazil
b) Srilanka
c) India
d) USA

958. India accounts for around ——% of world tea production

a) 23
b) 28
c) 18
d) 33

959. CIMMYT was established in

a) 1966
b) 1972
c) 1960
d) 1976

960 The world's largest producer of fruits is

a) India
b) Brazil
c) China
d) USA

961. The largest producer of milk in the world is

a) USA
b) Japan
c) China
d) India

962. Seed rate of oats is-

a) 40-50 kg/ha
b) 20-25 kg/ha
c) 15-20 kg/ha
d) 80-90 kg/ha

963. The kisan credit card scheme (KCCS) was introduced in

a) 1997 – 98
b) 1998 – 99
c) 1999 – 00
d) 2000 – 01

964. The National Agricultural Insurance Scheme (NAIS) was introduced in the country from the

a) Rabi 1999 – 2000
b) Kharif 1999 – 2000
c) Rabi 2000 – 2001
d) Kharif 2000 –2001

965. Pernicious anaemia occurs due to the deficiency of

a) Vitamin C
b) Vitamin B1
c) Vitamin B_{12}
c) Vitamin A

966. The millet produced in the largest quantity in the country is
   a) Pearl millet b) Maize
   c) Sorghum d) Finger millet

967. Henbane is a
   a) Qualitative SDP b) Quantitative SDP
   c) Qualitative LDP d) Quantitative LDP

968. NCIPM is situated at
   a) New Delhi b) Hyderbad
   c) Hyderbad d) Jhansi

969. The single most damaging pert of cotton accounting for almost half the total expenditure incurred by the farmer is
   a) Aphid b) Bollworm
   c) Jassid d) Thrip

970. In which country transgenic cotton occupies more than ¾ of the total cotton
   a) USA b) India
   c) China d) Australlia

971. The *Handbook of Agriculture* is a publication of:
   a) IARI b) ICAR
   c) Ministry of Agriculture d) CRIDA

972. The first edition of the *Handbook of Agriculture* was brought out in
   a) 1951 b) 1961
   c) 1971 d) 1981

973. The 'arable land' would comprises the
   a) Net area sown b) Current fallows
   c) Other fallow land d) *All of these*

974. The interveinal chlorosis of younger leaves is due to deficiency of
   a) Fe b) Mg
   c) S d) Zn

975. The micro – organisms which obtain their food by brcaking down dead plant and animal remains are called
   a) Parasites b) Saprophytes
   c) Pathogens d) None of these

976. The blast of rice is caused by

a) *Helminthosponum oryzae* b) *Pyricularia oryzae*

c) *Gibberella feykuroi* d) Neovossia horrid

977. The karnal bunt of wheat is caused by

a) *Meovossia indica* b) *Urocystis tcitici*

c) *Ustilago tritici* d) None of these

978. The kodo millet is:

a) *Setaria italica* b) *Panicum miliaceum*

c) *Paspalum scrobiculatum* d) *Eleusine coracanad*

979. The white nest of crucifers is caused by

a) *Melampsora lini* b) *Fusarium lini*

c) *Alaugo candida* d) None of these

980. The Rice hispa is

a) *Pachydiphosis* b) *Dicladispa*

c) *Gytocorisa* d) *Tryporyza*

981. The mustard sawfly is

a) *Bagrada* b) *Athalia*

c) *Plutella* d) Spodoptera

982. The ear – cockle disease of wheat is caused by

a) *Heterodera avenae* b) *Anguina tritici*

c) *Corynebacterium tritici* d) *Pratylenchus thornei*

983. The central Institute of Agricultural Engineering (CIAE) is located at

a) Ludhiana b) Nagpur

c) Kolkota d) Kanpur

984. Root pressure is measured by

a) Pressure bomb b) Osmometer

c) Potometer d) Manometer

985. The most important commodity presently placed in cold stores is

a) Potato b) Onion

c) Sea – foods d) Fruits

986. The Net capital ratio is given by

a) Deferred liabilities / Net worth

b) Total assets / Total liabilities

c) Arors income / Total assets × 100

d) Current assets / Current liabilities

987. The National Developments were first initiated in

a) 1960 b) 1965

c) 1970 d) 1975

988. The community Development projects were started in different parts of the country on

a) October 2, 1952 b) October 2, 1953

c) October 2, 1959 d) April 24, 1993

989. The Intensive Agricultural District Programme (Package programme) was started in

a) 1960 b) 1964

c) 1968 d) 1972

990. The minimum wages Act was enacted by the government of India in

a) 1948 b) 1949

c) 1950 d) 1951

991. The Insectirides Act was enacted by the government of India in

a) 1966 b) 1968

c) 1970 d) 1972

992. The pulse which is fawourly called as winter pulse is

a) Chick pea b) Black gram

c) Horse gram d) Red gram

993. The Dough stage in wheat is

a) Stage between milking and ripening b) Ripened stage

c) Early tillecing stage d) Heading stage

994. The micronutrient associated with pollen germination is

a) K b) B

c) Cu d) Zn

995. When mean = variance then the distribution is

a) Normal distribution
b) Poisson distribution
c) Binomial distribution
d) Random distribution

996. The basic unit of rural society is

a) Village
b) Panchayat Raj
c) Panchayat Samiti
d) Temple

997. The gene bank of India is located at:

a) NBPGR
b) IARI
c) ICAR
d) Botanical , Garden, Kolkata

998. One which is most suitable for study of mutations is

a) Haploid
b) Diploid
c) Tetraploid
d) Plyploid

999. If the number of treatments are 5 then error degree of freedom in *latin square design* will be

a) 16
b) 12
c) 8
d) 4

1000. The *Handbook of Horticulture* is published by

a) Horticulture society of India
b) ICAR
c) IARI
d) Ministry of Agricultured)

1001. More than _____ per cent of the earth's surface are covered with water

a) 55
b) 70
c) 76
d) 80

1002. Young plants contain _____________% water of their body weight; however, mature plants contain much less water than this.

a) 55-65
b) 70-75
c) 85-90
d) None of these

1003. In semi-arid regions, availability of rainfall / annum is ______ mm.

a) 200-400
b) 250-500
c) 300-550
d) 600-700

1004. On an average, ______ days is wet days in a year in our country (India).

a) 130
b) 145
c) 155
d) None of these

1005. ________________ irrigation method is best for undulated topography

a) Check basin
b) Low irrigation
c) Sprinkler
d) Furrow irrigation

1006. Which of the following criteria of irrigation scheduling is more acceptable to the farmers?

a) Soil moisture tension
b) Pan evaporation
c) Water potential of plants
d) Critical growth stages

1007. WUE is highest in

a) Flow irrigation
b) Sprinkler irrigation
c) Drip irrigation
d) All of these

1008. Most critical stage for irrigation in wheat is

a) Crown root initiation
b) Tillering
c) Flowering
d) Dough

1009. Sprinkler irrigation is ideal for which crop?

a) Tea
b) Jute
c) Mustard
d) Wheat

1010 Root depth of potato considered for irrigation is

a) 15 cm
b) 30 cm
c) 45 cm
d) 60 cm

1011. __________ ranks first in area and second in production of rice in the

a) China
b) India
c) Thiland
d) Indonesia

1012. The yield of *boro rice* is more than *kharif* (*aman*) rice is mainly due to ________________.

a) Cloudy days and poor water management
b) Sunny days and better water management
c) Cloudy days and poor nutrient management
d) Sunny days and poor nutrient management

1013. Green fodder of oat contains________% protein.

a) 5-7
b) 15-20
c) 10-12
d) 20-23

1014. ____________________________ is a variety of Maize.

a) Kent b) African Tall

c) Mescavi d) Panna

1015. The average green forage yield of Berseem is______________ t/ha.

a) 80-100 b) 110-115

c) 150-180 d) 160-200

1016. *Avena sativa* is the scientific name of

a) Maize b) Oat

c) Lucerne d) Barley

1017. Seed rate of Lucerne is ______________ kg/ha.

a) 10-12 b) 15-18

c) 20-24 d) 25-28

1018. Which of the following is a variety of Lucerne?

a) Anand-2 b) OS-6

c) Jho-822 d) P 3396

1019. Berseem is believed to be indigenous to

a) Egypt b) Europe

c) Ethiopia d) India

1020. Seed rate of fodder maize is__________________ kg/ha.

a) 50 b) 60

c) 70 d) 40

1021. *Azolla* is used as biofertilizer as it has

a) Rhizobium b) Cyanobacteria

c) Mycorrhiza d) Large quantity of humus

1022. Encourage natural cycle in organic farming by growing ____ crop.

a) Legume b) Cereals

c) Millets d) Oilseed

1023. Conversion from NO_2^- to NO_3^- is carried out by

a) 3 b) 5

c) 8 d) 1

1024. Biofertilizer supporting Sorghum is

a) Rhizobium b) Azospirillum

c) Azotobactor d) Anabaena

1025. Conversion period for fully certified organic is________ years.

a) 3 b) 5

c) 8 d) 8

1026. ___________ is a Biopesticide

d) Bt e) PSB

f) VAM d) Acetobactor

1027. To manage crop pest by growing non-host crop

a) Trap crop b) Buffer crop

c) Cover crop d) Decoy crop

1028. *Subabul is* an example of

a) Green leaf manure b) Green manure

c) Cover crop d) Mulch crop

1029. Certified Organic logo in India –

a) India Organic b) India Green

c) India Healthy d) India Fresh

1030. The Pest control property in *neem* product is due to ___________

a) Azadirectin b) Nicotine

c) Daturin d) Thine

1031. Scientific name of mulberry silk worm is

a) Bombyx mori b) Philosamia ricini

c) Antheria assami d) None of these

1032. Usual sowing time of lentil in eastern part of India is

a) Mid Oct. to Mid Nov.

b) 3rd week of Nov. to 1st week of Dec.

c) 2nd fortnight of Dec.

d) None of these

1033. Which of the following crops has the highest cultivated area in the world?

a) Rice b) Wheat

c) Bajra d) Barley

1034. Rice inflorescence is known as

a) Ear b) Pod

c) Panicle d) Siliqua

1035. __________ crop contain highest protein

a) Urd bean b) Maize

c) Soybean d) Mung bean

1036. ICPL 87 (*Pragati*) is a variety of

a) Soybean b) Maize

c) Urdbean d) Pigeon pea

1037. Fruit of chickpea is known as

a) Greengram b) Blackgram

c) Redgram d) Bengal gram

1038. PDU 1 is a variety of

a) Greengram b) Blackgram

c) Redgram d) Bengal gram

1039. 'Indian farming' is published from

a) IARI b) NBPGR

c) ICAR d) IASRI

1040. PL 406 is a variety of –

a) Chickpea b) Lentil

c) Lathyrus d) Mothbean

1041. Primary centre of origin of blackgram is

a) Central Asia b) China

c) India d) Africa

1042. The word 'Agriculture' is derived from two words, 'agric' means ____________ and 'cultura' means cultivation

a) Field b) Soil

c) Land d) None of these

1043. One acre is equivalent to __________ ha.

a) 0.603 b) 0.04

c) 1.30 d) 0.404

1044. Duck foot tyne is mainly used ________

a) to open furrow
b) to dig soil
c) to control weeds
d) None of these

1045. 'Agronomy' is derived from two _________ words, 'agros' and 'nomos'.

a) Latin
b) Greek
b) Arabic
d) Sanskrit

1046. Sickle is an ___________________

a) Machine)
b) Implement
c) Tool
d) None of these

1047. The ideal tillage method for semi-arid tropic region is ______________.

a) Repeated deep tillage
b) Blind tillage
c) Zero tillage
d) Reduced tillage

1048. Which of the following is not a preparatory tillage

a) Ploughing
b) Harrowing
c) Hoeing
d) Planking

1049. Layout of crop field is _________.

a) Primary tillage
b) Inter cultivation
c) Preparatory tillage
d) All of these

1050. Soil physical property affected by tillage operation -

a) Bulk density
b) Soil moisture
c) Soil colour
d) All of these

1051. The angle between two arms of the country plough is

a) 85°
b) 100°
c) 115°
d) 135°

1052. Mulch is used to reduce

a) Run-off flow
b) Evaporation
c) Seepage
d) None of these

1053. Edaphic micro-environment to the crops is provided by

a) Irrigation
b) Mulching
c) Drainage
d) None of these

1054. Residue management is very common practices in developed countries to leave_______________at the field after crop harvest

a) Straw
b) Stubble
c) Green portion
d) None of these

1055. Farm ponds are generally constructed to provide

a) Irrigation water
b) Life-saving irrigation
c) Fish cultivation
d) None of these

1056. _________________is the way of effective utilization of residual soil moisture and nutrients

a) Paira cropping
b) Relay Cropping
c) Mixed cropping
d) None of these

1057. Fixing up of time interval of water application is known as ___________.

a) Irrigation measurement
b) Irrigation scheduling
c) Irrigation application
d) None of these

1058. Calculation of soil moisture requires_________.

a) bulk density
b) particle density
c) transpiration rate
d) None of these

1059. First genetically modified field crop is_______

a) Cotton
b) Wheat
c) Maize
d) Rice

1060 Rice belongs to family

a) Asteraceae
b) Poaceae
c) Fabaceae
d) Brassicaceae

1061. Crop rich in source of carotene is

a) Carrot
b) Rice
c) Potato
d) Wheat

1062. Soil moisture is measured using _________________

a) Barometer
b) Tensiometer
c) Lactometer
d) Lycimeter

1063. Water use efficiency is expressed in_______.

a) mm
b) g/cm
c) kg/ha/cm
d) kg/cm

1064. *Parthenium hysterophorus* is an

a) Invasive weed
b) Indigenous weed
c) Invasive alien weed
d) Crop plant

1065. Glyphosate 71 SG is a

a) Selective herbicide
b) Non-selective herbicide
c) Safer herbicide
d) Both b&c

1066. Collego is a

a) Mycoherbicide
b) Chemical herbicide
c) Both
d) None of these

1067. Paraquat dichloride 24 SL is used to kill all types of weeds as POE in________.

a) Crop field
b) Roadside & Waste land
c) Both
d) None of these

1068. *Eicchornia crassipes* is an

a) Indigenous weed
b) Annual weed
c) Aquatic weed
d) Jute field weed

1069. Scientific name of Witch weed is

a) *Leucas aspera*
b) *Striga asiatica*
c) *Vicia sativa*
d) *Parthenium hysterophorus*

1070. *Cuscuta reflexa* is commonly known as______________

a) Jimson weed
b) Mutha gash
c) Dodder
d) Horse weed

1071. Safener is used with herbicides to

a) Protect the crop
b) Increase activity
c) Increase phytotoxicity
d) None of these

1072. Acetamide herbicide Butachlor is commonly used in ________.

a) Cotton
b) Tea
c) Paddy
d) Jute

1073. *Cyperus rotundus* is a

a) Perennial weed
b) Sedge weed
c) Both
d) None

1074. Citronella is used for

a) Soap b) Cosmetics

c) Perfumery d) All of these

1075. Seed rate of *Sarpagandha* is

a) 5-6 kg/ha b) 10-12 kg/ha

c) 15-16 kg/ha d) 20-25 kg/ha

1076. Groundnut seed is known as _______________

a) Pod b) Caryopsis

c) Siliqua d) Kernel

1077. Linseed can be used as _______________ crop

a) oilseed b) fibre

c) oilseed and fibre d) pulse

1078. "Recinine" an alkaloid generally found in _______________

a) sesame b) niger

c) safflower d) castor

1079. Rancidity in sunflower oil is caused by the process of _______________

a) reduction b) oxidation

c) esterification d) nitrification

1080. Which among the followings is termed as golden gram?

a) Green gram b) Black gram

c) Horse gram d) Redgram

1081. FIRB technology is planting of _______________

a) Wheat b) Pearl millet

c) Finger millet d) Proso millet

1082. What is commonly used soil:water ratio for pH determination

a) 1:1 b) 1:2

c) 1:1.5 d) 1:2.5

1083. Element essential for the deflocculation of soil particles ___________

a) K b) Ca

c) Mg d) Na

1084. *Eleucine corocana* is the botanical name of _______________

a) foxtail millet b) pearl millet

c) finger millet d) proso millet

1085. Which form of phosphorus is water soluble?

a) Monovalent b) Divalent

c) Trivalent d) None of these

1086. Nitrogen use efficiency is comparatively lower in _______________

a) Wheat b) Maize

c) Paddy d) Chickpea

1087. National Research Centre for Soybean ia located at _______________

a) Indore b) Bangalore

c) New Delhi d) Junagarh

1088. Macropores of the soil are usually filled with _______________

a) Water b) Air

c) Both d) None

1089. Rock phosphates are generally used in soils which are _______________

a) strongly acidic b) strongly alkaline

c) neutral soils d) saline soils

1090. Isolation distance for foundation seed production of rice is

a) 3 m b) 30 m

c) 50 m d) 100 m

1091. Rotational intensity of Rice-wheat crop rotation is _______________

a) 100% b) 200%

c) 300% d) 400%

1092. If maize is sown at spacing of 60×25cm then plant population of maize in 6000 m^2 area will be _______________

a) 10000 b) 20000

c) 30000 d) 40000

1093. Sustainability yield index (SYI) lies between _______________

a) 0-1 b) -1 to +1

c) Above 1 d) None of these

1094. Most common tillage implement used in india is _______________

a) Country plough
b) MB plough
c) Victory plough
d) Harrows

1095. The hardening of surface soil is technically termed as _______________

a) Compaction
b) Pulverization
c) Hard pan
d) None of these

1096. For the correction of boron deficiency borate is applied to soil @ ________

a) 1-2 kg/ha
b) 2-3 kg/ha
c) 3-4 kg/ha
d) 4-5 kg/ha

1097. Crop recovery efficiencies for fertilizer P is

a) 10-20%
b) 20-30%
c) 10-15%
d) 20-25%

1098. Crop recovery efficiencies for fertilizer K is

a) 10-20%
b) 20-30%
c) 30-40%
d) 40-50%

1099. Which one of the following is moderately mobile element in plant body

a) N
b) P
c) Zn
d) B

1100. Biological fertility of soil is _______________ in nature.

a) Static
b) Dynamic
c) Either a or b
d) None

1101. VAM is a ____________________

a) Virus
b) Bacteria
c) Fungi
d) Bacteria

1102. __________________ irrigation method is best for undulated topography

a) Check basin
b) Flow irrigation
c) Sprinkler
d) Furrow irrigation

1103. On an average, ________ days is wet days in a year in India

a) 130
b) 140
c) 150
d) None of these

1104. Most critical stage for irrigation in wheat is ___________

a) CRI
b) Tillering
c) Flowering
d) Dough

1105. The average green forage yield of Berseem is_______________ t/ha

a) 80- 100
b) 110-150
c) 150-180
d) 160-200

1106. Seed rate of Lucerne is _____________ kg/ha

a) 10-12
b) 15-18
c) 20-24
d) 25-28

1107. *Azolla* is used as biofertilizer as it has ___________________

a) Azotobacter
b) Cyanobacteria
c) Mycorrhiza
d) Large quantity of humus

1108. Berseem is believed to be indigenous to _______________

a) Egypt
b) Europe
c) Ethiopia
d) India

1109. ___________ is a Bio-pesticide

a) Bt
b) PSB
c) VAM
d) *Azolla*

1110 The crop which has the highest cultivated area in the world is________

a) Rice
b) Wheat
c) Bajra
d) Barley

1111. Certified Organic logo in India

a) India Organic
b) India Green
c) India Healthy
d) India Fresh

1112. Hilum of green gram seed is ______________

a) Concave
b) Convex
c) Flat
d) Round

1113. 'Indian farming' is published from ________________

a) IARI
b) IIPR
c) ICAR
d) IASRI

1114. A short duration crop in between two main crops is known as

a) Companion crop b) Cash crop

c) Catch crop d) Intercrop

1115. One acre is equivalent to __________ ha

a) 0.603 b) 0.04

c) 1.30 d) 0.404

1116. Spade is an ____________________

a) Machine b) Implement

c) Tool d) None of these

1117. Layout of crop field is a part of _________.

a) Primary tillage b) Interculture

c) Preparatory tillage d) None of these

1118. There are _____________ cultivated species of *Oryza*

a) 8 b) 6

c) 4 d) 2

1119. Mulch is used to reduce ________________ loss of water

a) Percolation b) Evaporation

c) Seepage d) None

1120. First genetically modified field crop is_______

a) cotton b) wheat

c) maize d) brinjal

1121. PSB increases the availability of __________

a) N b) S

c) K d) P

1122. ***Satabdi*** is a good variety of __________________

a) Wheat b) Barley

c) Maize d) Rice

1123. *Parthenium hysterophorus* is an _______________

a) Invasive weed b) indigenous weed

c) invasive alien weed d) *indoor plant*

1124. __________ creates an impervious layer which help to reduce percolation of water

a) Planking b) Puddling

c) Spading d) Hoeing

1125. Percentage of edible oil in soybean is____________

a) 10-12% b) 13-15%

c) 16-18% d) 20-22%

1126. Collego is a _____________________

a) Mycoherbicide b) Chemical herbicide

c) Both d) None

1127. The cereal non-tillering in nature is ________________

a) Rice b) Wheat

c) Maize d) Barley

1128. Citronella is used for the preparation of _____________

a) Soap b) Cosmetics

c) Perfume d) All

1129. *Sonali,* and *Samrat* are two improved varieties of ______________

a) black gram b) lentil

c) bengal gram d) green gram

1130. The fruit of lentil is known as ________________

a) caryopsis b) pod

c) grain d) siliqua

1131. Linseed crop can be used as___________________

a) oilseed b) fibre

c) both purpose d) pulse

1132. The inflorescence of sugarcane is called as ________

a) arrow b) tassel

c) panicle d) spike

1133. *Avena sativa* is the scientific name of __________

a) Maize b) Oat

c) Lucern d) Barley

1134. Central Rice Research Institute is located at ___________

a) Hyderabad b) Kolkata

c) Guntur d) Cuttuk

1135. The value of correlation coefficient lies in between ___________

a) -2 to 0 b) 0 to +1

c) -1 to +1 d) 1 to 2

1136. The idea of randomization was put forwarded by ___________

a) Chapman b) Gomez

c) Sukathmae d) Fisher

1137. Poor man's timber is ___________

a) Mango wood b) Jackfruit wood

c) Babla wood d) Bamboo

1138. The crop considered as queen of cereal crops is ___________

a) Oat b) Maize

c) Rice d) Wheat

1139. Hybrid rice cultivation is very popular in ___________

a) India b) Japan

c) Phillipines d) China

1140. An example of narcotic crop is ___________

a) Indian hemp b) Opium poppy

c) Tobacco d) All of these

1141. The important nutrients for oilseed crops are

a) N & P b) P &K

c) N, P & K d) N, P, K & S

1142. **S** content of SSP (Single Super Phosphate) is ___________

a) 8% b) 10%

c) 12% d) 16%

1143. The entire process from rough rice to head rice is considered to determine ___________ quality

a) Milling b) Eating

c) Nutritional d) Hulling

1144. Soil solarisation is done to destroy __________

a) Pathogen b) Insect pest

c) Weed seed d) All of these

1145. *Cyperus rotundus* is a

a) Grass weed b) Sedge weed

c) Perennial weed d) Both b& c

1146. MPT refers to __________

a) Multi-production tree species b) Multi-purpose tree species

c) Maximum productive species d) None of these

1147. Example of concentrated organic manure __________

a) Oilcake b) FYM

c) Fish meal d) Both a& c

1148. Primary centre of origin of black gram is __________

a) Central Asia b) China

c) India d) Africa

1149. The pest control property of *neem* is due to presence of __________

a) Daturin b) Thine

c) Nicotine d) Azadiractin

1150. The major chemical compound for aroma in rice is

a) 2-acctyl-1-pyrroline

b) 2-acctyl-1pyrazine

c) 2-acctyl 2 thiazoline

d) 2-acctyl-1,4,5,6 tetra hydro pyridine

1151. The quantity of urea needed for applying 115 kg Nitrogen will be

a) 110 kg b) 185 kg

c) 215 kg d) 250 kg

1152. Application of nitrogen in pulses at the time of planting is known as

a) Basicdose b) Starterdose

c) Additionaldose d) Synergistic dose

1153. The highest nitrogen content among the commonly used nitrogenous fertilizers is embodied by

a) Ammonium nitrate b) Ammonium sulphate

c) Calciumammonium nitrate d) Urea

1154. Nitrification is a __________ process

a) physical b) chemical

c) biological d) physico-chemical

1155. Rock phosphate can be directly applied as fertilizer in __________ soils

a) acidic b) alkaline

c) neutral d) calcareous

1156. Symbiotic association of fungal hyphae with plant roots is known as

a) lichen b) co-parasitism

c) mutual infection d) mycorrhiza

1157. Inherent capacity of soil to supply nutrients in adequate amount and suitable proportions is called as

a) fertility index b) productivity index

c) soil fertility d) soil productivity

1158. Among the following, acid forming fertilizer is __________

a) Urea b) Sodium nitrate

c) Potassium chloride d) CAN

1159. Maximum amount of P_2O_5 is in __________

a) DAP b) DCP

c) SSP d) TSP

1160 Potassium salts are...

a) more mobile than nitrates b) less mobile than P

c) less mobile than nitrates d) as mobile as nitrates

1161. The best green manuring crop for garden land is __________

a) guar b) dhaincha

c) sunnhemp d) berseem

1162. Split doses of nitrogen is used in __________ type of soil

a) sandy b) saline

c) clayey d) alkaline

1163. Urea is a __________ fertilizer

a) inorganic b) organic

c) ammonical d) nitrate

1164. Which of the following contains most of the plant nutrients?

a) Urea b) FYM

c) MOP d) Rock phosphste

1165. *Khaira* disease in rice is caused due to __________

a) Zinc deficiency b) fungal infection

c) bacterial infection d) nitrogen deficiency

1166. Fertilizer which supplies three essential plant nutrients is

a) DAP b) SSP

c) MOP d) SOP

1167. *Akiochi* disease in rice occurs due to the toxicity caused by __________

a) Zn b) B

c) P d) H_2S

1168. Urea is a __________ fertilizer

a) Straight b) Complex

c) Mixed d) Compound

1169. The N deficiency symptoms occur in __________

a) upper leaves b) lower leaves

c) middle leaves d) topmost bud

1170. Available K less than 100kg ha^{-1} in soil is __________

a) Low b) Medium

c) high d) neutral

1171. Concentrated organic manure

a) Oilcake b) FYM

c) Vermicompost d) All of these

1172. Encourage natural cycle in organic farming by growing

a) cereals b) millets

c) legumes d) none of these

1173. For boro rice, ___________ N should be given as basal.

a) more b) less

c) zero d) total

1174. Neem cake contains ___________ % N.

a) 5 b) 6

c) 7 d) 8

1175. Conversion period for fully certified organic is

a) 1 year b) 2 years

c) 3 years d) 4 years

1176. Name of a bio-pesticide is

a) PSB b) VAM

c) NPV d) BGA

1177. *Glyricidia* is an example of

a) green manure b) green leaf manure

c) mulch crop d) all of these

1178. Crop pest can be managed by growing ___________ crop

a) GM b) cover

c) trap d) nurse

1179. IFOAM was established in the country ___________

a) India b) China

c) France d) Russia

1180. The pest control property of tobacco is due to presence of

a) Daturin b) Thine

c) Nicotine d) Azadirectin

1181. LEISA is

(a) Low External Input and Sustainable Agro-forestry

(b) Low-External Input and Sustainable Agriculture

(c) Low Energy and Input Saving Agriculture

(d) Local Export and Import Selling Authority

1182. In India, the organic quality control has been implemented by the Government of India under

a) NPOF b) HACCP

c) NPOP d) FAO

1183. Which of the following is called as "Farmers' friend"?

a) VAM b) PSB

c) Azolla d) Earthworm

1184. An organism that captures and eats another organism, are called

a) Parasite b) Predator

c) Pathogen d) Omnivore

1185. Which of the following is not true about vermicompost

a) It is rich in nutrients

b) it is easily available to plants

c) It possess some enzymes and growth promoting substances

d) It has long term hazardous effect on crops.

1186. Which is of the following is not a component of organic agriculture

a) Organic manures

b) Farm machinary

c) Non-chemical weed control

d) Biological pest and disease management

1187. Which of the following is not an organic agriculture

a) Green manuring

b) Brown manuring

c) Intensive cultivation using hybrids

d) Natural farming

1188. The continent with most organic land is in

a) Asia b) Europe

c) Australlia d) South America

1189. Conversion period in organic certification for annuals crops is

a) 1 year b) 2 years

c) 3 years d) 4 years

1190. Which one of the following organic manure has the narrowest value of C/N ratio?

a) Fern b) Groundnut cake

c) Biogas slurry d) FYM

1191. Biofertilizers are mainly applied in the fields with the main purpose of

a) Protection from pest
b) Sustainability
c) Reduce cost of fertilizer
d) Increase return

1192. National Centre for Organic Farming is situated at

a) Ghaziabad, U.P.
b) New Delhi
c) Coimbatore, Tamil Nadu
d) Ranchi, Jharkhand

1193. HACCP denotes

a) Hazard Analysis Critical Control Points
b) Human Analysis Chemical Cultivation Process
c) Human Assurance Critical Control points
d) Hazard Analysis Critical Chemical Production

1194. Which one of the following is the principal N- fixing algal community of blue-green algae?

a) Rhizobia
b) Anabaena
c) Azotobacter
d) Azospirillium

1195. Which of the following oil-cakes, in addition to its manurial value, also acts as a nitrification inhibitor?

a) Mustard
b) Groundnut
c) Neem
d) Sesame

1196. Mobility of S in plant body is very ___________

a) high
b) less
c) either a or b
d) neither a nor b

1197. C_3 plants assimilate nitrate efficiently than C_4 plants.

a) more
b) less
c) either a or b
d) neither a nor b

1198. GMO crop is .... in organic farming

a) encouraged
b) prohibited
c) restricted
d) popularly cultivated

1199. For good yield rice crop needs at least ___________ continuous sunny days after heading

a) 15
b) 20
c) 25
d) 30

1200. The opium is harvested when the plants attain ___________ maturity

a) physiological b) industrial

c) harvest d) delayed

1201. The word 'Agriculture' is derived from two words, 'agric' means ___________ and cultura means cultivation.

a) field b) soil

c) land d) crop

1202. Duck foot tyne is mainly used for

a) to open furrow b) to dig soil

c) to control weeds d) None of these

1203. 'Agronomy' is derived from two ________ words, 'agros' and 'nomos'

a) Latin b) Greek

c) Arabic d) Sanskrit

1204. Sickle is an ___________

a) Machine b) Implement

c) Tool d) None of these

1205. The ideal tillage method for semi-arid tropic region is ___________

a) Repeated deep tillage b) Blind tillage

c) Zero tillage d) Reduced tillage

1206. Which of the following is not a preparatory tillage ___________

a) Ploughing b) Harrowing

c) Hoeing d) Planking

1207. Layout of seed bed is ___________

a) Primary tillage b) Inter cultivation

c) Preparotory tillage d) All of these

1208. Agricultural operation requiring maximum non-biological power is

a) Harrowing b) Ploughing

c) Harvesting d) Spraying

1209. Which of the following soil physical property is affected by tillage ?

a) Bulk density b) Soil moisture

c) Soil colour d) All of these

1210. The angle between two arms of the country plough is ___________

a) 85º b) 100º

c) 115º d) 135º

1211. Mulching is used to reduce ___________

a) run-off flow b) percolation

c) seepage d) infiltration

1212. Edaphic micro-environment to the crops is provided by

a) Irrigation b) Mulching

c) Drainage d) Fertilizer

1213. Residue management is very common practices in developed countries to leave ___________ at the field after crop harvest.

a) straw b) stubble

c) green portion d) both a and c

1214. Irrigation is application of ___________ to the crop field.

a) water b) supplemental water

c) complemental water d) additional water

1215. Farm ponds are generally constructed to provide

a) irrigation water b) life saving irrigation

c) fish cultivation d) all of these

1216. Water use efficiency is expressed

a) mm/kg b) g/cm

c) kg/ha/cm d) g/ha/mm

1217. Impounded Farm pond is suitable in

a) Hilly b) Coastal

c) Alluvial region d) Laterite belt

1218. ___________ is the way of effective utilization of residual soil moisture and nutrients.

a) Paira Cropping b) Relay Cropping

c) Mixed Cropping d) Intercropping

1219. Irrigation is ___________ water application in crop cultivation when crop needs.

a) additional b) supplemental

c) incremental d) complemental

1220. CRI is ___________ growth stage of wheat crop comes after 21 days of sowing.

a) physiological
b) critical
c) morphological
d) morpho-physiological

1221. Water requirement of crop usually comes from ___________ which commonly known as consumptive use of water.

a) ET
b) percolation
c) seepage
d) infiltration

1222. To fix up time interval of water application is known as

a) irrigation measurement
b) irrigation scheduling
c) irrigation application
d) None of these

1223. Calculation of soil moisture requires

a) Bulk density
b) Particle density
c) Transpiration rate
d) All of these

1224. First genetically modified field crop is

a) Cotton
b) Wheat
c) Maize
d) Tomato

1225. Soil moisture is measured using

a) Barometer
b) Tensiometer
c) Lactometer
d) Water meter

1226. Which of the tree species is preferred under heavy soils <800 mm rainfall

a) Sesbania grandiflora
b) *Leucaena leucocephala*
c) *Glyricidia sepium*
d) *Cassia siamiea*

1227. Most common agro forestry in arid and semiarid region is

a) Agro silviculture
b) silvipasture
c) Horti silvipasture
d) Agri horti silviculture

1228. Best tree species suitable under wetland condition is

a) *Taxodium scadens*
b) *Salix spp*
c) *Alinus trabeculosa*
d) *Morus alba*

1229. Which of the following tree species are tolerant under water logged condition and grow fast and have good timber quality

a) *Taxodium distichum* (Sovamp cypress)

b) Water fir

c) Peduncled

d) All

1230. Social forestry means:

a) It is forestry of people by the people and for the people

b) The forestry in which the efforts aimed at raising and managing trees for the benefits of rural People

c) Both a & b

d) None

1231. Essential ingredients for the success of community forestry are

a) Capabilities of the land

b) The villager's choice of land use

c) Nature of the support the organizational structure provides

d) All

1232. According to Jeff Romm (1980) which is essentials for the community forestry programme are

a) The technology must be more suitable and available and must be more productive and ecologically sustainable than that already in use

b) The villagers must feel secured of the benefits

c) New uses of the land and other resources must be profitable from the villager's point of view

d) All

1233. The main problems faced in implementing community forestry are:

a) Villagers feel apprehensive of the programme, which they feel is another version of taking away their land by the government

b) Programme needs community's voluntary cooperation and cannot be imposed on it

c) Programme should be based on the trust and confidence on both sides

d) All

1234. Urban forestry is:

a) It is specialized branch of forestry that has as its objective the cultivation and management of trees for their present

b) Potential contribution to the physiological, socio-logical and economic well being of urban society

c) Both a & b

d) None

1235. Management of Urban Forest is concerned with management needs of forests in urban areas and with how needs are:

a) Whatever is done for the forest to maintain the health and vigour

b) Whatever is done to the forest to prevent undue interference with the society

c) Both a & b

d) None

1236. Benefits of Urban Forests are:

a) Climate amelioration

b) Environmental engineering uses Architectural uses

c) Aesthetic uses

d) All of these

1237. Use oriented forestry are:

a) Industrial forestry　　b) Energy forestry

c) Both a & b　　d) None

1238. Benefits of social forestry are:

a) Betterment of environment

b) Reduction of pollution

c) Protection from wind, conservation of moisture

d) All

1239. The history of Agro forestry is:

a) Cultivating trees and agricultural crops in intimate combination with one another is an ancient practice that farmers have used throughout the world

b) Until the middle Ages, it was the general custom to clear-fell degraded forest, burn the slash, cultivate food crops for varying periods on the cleared areas

c) Plant or sow trees before, along with or after sowing agricultural crops

d) All

1240. 'Agroforestry' the definition implies that:

a) Agro forestry normally involves two or more species of plants at least one of which is a woody perennial

b) An agro forestry system always has two or more outputs

c) The cycle of an agro forestry system is always more than one year

d) All

1241. Theoretically, all agro forestry system posses are:

a) Productivity
b) Sustainability
c) Adoptability
d) All

1242. Agro forestry systems can be categorized to sets of criteria:

a) Structural basis
b) Functional basis
c) Socio-economic basis
d) Ecological basis

1243. The framework for classification of Agro forestry System and practices are:

a) Structure of the system (nature and arrangement of components)

b) Function of the system (role and output of components)

c) Socio-economic seals and management levels of the system

d) All

1244. What are MPTs?

a) The multipurpose tree and shrubs could be defined as trees grown deliberately

b) Kept and managed for preferably more than one intended use

c) Services in any multipurpose land use system, especially agroforestry system

d) All

1245. The multiple benefits of MPTs are:

a) The trees with many uses would be preferred to those with one single use

b) But in a broad sense every tree has at least the triple functions of protective, productive and socio-economic roles

c) Both

d) None

1246. Protective (environment) is

a) Moderation of micro/macro climatic parameters, checking soil erosion and water run-off

b) Soil improvement/fertility build up, water conservation and flow moderation

c) Wildlife habitats

d) All

1247. Productive means

a) Wood: timber, building material, veneers, chipboards and other panel products, pulp and paper, rayon etc.

b) Bark: Raw as fuel, dyes, tennis and chemical extraction etc

c) Both

d) None

1248. Energy means

a) Raw: Firewood

b) Processed: Charcoal, gases or liquid fuels, chemical stem extractives, resin, oil, paint, varnishes

c) Leaf: Thatch, fodder, oil, silk, wrapper, medicines, honey, dyes, food

d) Root: Fiber, fuel wood, dyes, and chemical extractives

1249. Socio-Economic includes

a) Employment generation

b) Income generation - foreign exchange and import substitution

c) Public education

d) All

1250. Which of the following is the correct statement?

a) Principle of agri-silviculture is similar to that of intercropping

b) Principle of intensive cropping

c) Principle of agri-silviculture similar to that of double cropping

d) Principle for agri-silviculture is similar to that of mono-cropping

1251. What is agroforestry?

a) Agro forestry is the integration of agriculture and/or farming with forestry so the land can simultaneously be used for one purpose

b) Agro forestry is the integration of agriculture and/or farming with forestry so the land can simultaneously be used for more than one purpose

c) Either 'a' is true or 'b' is true

d) Nether 'a' is true nor 'b' is true

1252. The purpose of agroforestry are

a) The purpose of agroforestry is to integration of trees in farming systems and their management in rural landscapes to enhance productivity, profitability, diversity and ecosystem sustainability

b) The purpose of agroforestry is to manage in rural landscapes to enhance productivity, profitability, diversity and ecosystem sustainability.

c) Both 'a' & 'b'

d) None of the items 'a' & 'b' is true

1253. Agroforestry systems may be classified based on the following criteria

a) On structural & functional basis b) On a economical basis

c) On a socioeconomic basis d) On an ecological basis

1254. Well managed agroforestry systems may be provide

a) Well managed Agroforestry lands may provide food, fodder and fiber

b) Well managed Agroforestry system may also provide cultural and household utility items

c) Both 'a' & 'b' is true

d) None

1255. Well managed agroforestry system on slopping land may protect the land from

a) Landslides b) Soil erosion

c) Both a & b d) None

1256. In point moisture conservation, agroforestry system may helps

a) for better infiltration water in soil

b) to reduce the velocity of run-off water

c) Both a & b

d) None

1257. The fundamental aim of the agro forestry extension activities are
   a) to take appropriate measures in available fallow land for the betterment of the local community
   b) to upgrade their realistic insight thought
   c) to make to grow their insight ability
   d) Both a & b

1258. Agroforestry is a land-use systems involving trees combined with crops and/or animals on the same unit of land. It includes
   a) It involves the interplay of socio-cultural values more than in most other land-use systems;
   b) It is structurally and functionally more complex than monoculture
   c) Both
   d) None

1259. The tree species used for agroforestry have
   a) higher C sequestration capacity than forest trees
   b) higher C sequestration capacity than crop and grass system
   c) Both a & b
   d) None

1260 Agroforestry is a better for land-use management. It includes
   a) Production of multiple outputs with protection of the resource base
   b) Places emphasis on the use of multiple indigenous trees and shrubs
   c) Both
   d) None

1261. Function of Agroforestry system is
   a) Protein bank
   b) Live-fence of fodder trees and hedges
   c) Trees and shrubs on pasture
   d) All

1262. Agroforestry systems is involving deliberate management of multipurpose trees and shrubs in intimate association with annual and perennial agricultural crops and livestock within the compounds of individual houses
   a) The whole tree- crop- animal units are being intensively managed by family labor.
   b) Home gardens can also be called as Multitier system or Multitier cropping.

c) Food production is primary function of most home gardens

d) All

1263. The species, which are required for Agroforestry system for raising home garden or kitchen garden, are

a) Woody species: Anacardium occidentale, Artocarpus heterophyllus, Citrus spp, Psiduim guajava, Mangifera indica, Azadirachta indica, Cocus nucifera

b) Herbaceous species: Bhendi, Onion, cabbage, Pumpkin, Sweet potato, Banana, Beans, etc.

c) Both

d) None

1264. The practice of Agroforestry has been adopted to demarcate the boundary of home stead land /fish pond get more yield from the practice of Agroforestry

a) In Agroforestry system various woody hedges, especially fast growing and coppicing fodder shrubs and trees are planted for the purpose of browse, mulch, green manure, soil conservation etc. The following species viz., Erythrina sp, Leucaena luecocephala, Sesbania grandiflora are generally used.

b) In Agroforestry system various honey (nector) producing trees frequently visited by honeybees are planted on the boundary of the Home stead land.

c) In Agroforestry system various trees and shrubs preferred by fish are planted on the boundary and around fish ponds. Tree leaves are used as feed for fish. The main role of this system is fish production and bund stabilization around fish ponds.

d) All

1265. Agroforestry is a part of

a) Farm forestry

b) Social forestry

c) Both

d) None

1266. Agroforestry is intermediate between

a) Horticultural species and horticultural species

b) Forest species and agricultural species

c) Agricultural species and agricultural species

d) Forest species and forest species

1267. Agroforestry is practiced on

a) On forest land b) Non forest land

c) On uncultivated waste land d) None

1268. What is the primary objectives agroforestry?

a) To increase the yield of agricultural crops kg/ha

b) To increase the yield of forest crops in M.T

c) Production of agricultural crops (live stock as well) are integrated with the production of perennial woody species on same land

a) None

1269. The of agroforestry in present context

a) To increase Forest cover outside the Forest areas

b) To increase agricultural cover in non agricultural field

c) Both a & b

d) None

1270. Development of agroforestry is required in present day context

a) To increase the supply of food with increasing population

b) For meeting the enhance need of greater demand of fodder, fuel & Timber

c) Both

d) None

1271. The practice of agroforestry has positive effect on soil health

a) It increases the organic matter by way of addition of leaf litter

b) More efficient nutrient cycling

c) Both

d) None

1272. The practice of agroforestry helps in changing the physical characteristics of soil by way of

a) The practice of agro forestry helps in enhancing water holding capacity of soils

b) The practice of agro forestry helps in changing the porosity of soil, soil density and activities of micro organisms in soil

c) Both

d) None

1273. The practice of agroforestry helps in conservation & influences on hydrological characteristics of soil

a) The practice of agro forestry helps in conservation soil from erosion

b) Losses of moisture from soil by way of evaporation will be reduced

c) More moisture will be infiltrated into soil

d) All

1274. The negative aspects of the practice of agroforestry are

a) Fast growing species demand high water and nutrients

b) Formation of nutrient depletion zone

c) Allelopathy effects, accumulation of toxic exudates

d) All

1275. The practice of agroforestry helps in modifies the microclimate by

a) Reduces the velocity of wind

b) Maintains the solar radiation and temperature of the area and improves the water table

c) Increase rainfall in the area due to orographic effect of tree during monsoon season

d) All

1276. The practice of agroforestry will helps in deriving the social benefit by way of

a) The practice of agro forestry will helps in the Improvement of rural living standards by way of getting sustained employ-ment

b) The practice of agro forestry will helps in the Improvement in nutrition and health due to increased quality and diversity of food outputs

c) Both

d) None

1277. What is the social benefit of agroforestry system?

a) Sustainable employment

b) Improvement of quality

c) Stabilization the upland community

d) All

1278. In which state, *Taungya* system of cultivation is popular?

a) Kerala b) West Bengal

c) UP d) All

1279. Which is the major cause of deforestation?

a) Shifting cultivation b) Cutting tree of fuel wood

c) Both d) None

1280. Forest area is the highest in __________ state.

a) UP b) MP

c) Orissa d) Arunachal Pradesh

1281. Characteristics of suitable trees under agroforestry system is

a) Legume tree b) Deep rooted

c) Wide spaced one d) All

1282. Contribution of forest products to the world Gross Domestic Product is

a) 10% b) 1%

c) 5% d) 20%

1283. Indian forest act was enacted during

a) 1927 b) 1912

c) 1952 d) 1992

1284. Composition of components including spatial, temporal and vertical basis is the agro forestry criteria of classification

a) Functional basis b) Structural basis

c) Socio – economic basis d) Ecological basis

1285. Percentage of forest cover is highest in

a) Andaman and Nicobar Islands b) Assam

c) Mizoram d) Tripura

1286. Major objective of the agroforestry is

a) Productivity b) Sustainability

c) Adaptability d) All

1287. Percentage of forest cover in the world to the area is

a) 25 b) 10

c) 50 d) None

1288. Continent has the highest area of forest is

a) Africa b) Asia

c) Europe d) South America

1289. Primary aim of alley cropping is

a) Increasing yield b) Improving soil health

c) Weed control d) All

1290. Main purpose of following shifting cultivation is

a) Restoring soil fertility b) Reduction erosion

c) Both d) None

1291. Tree species suitable for live-fence of fodder and hedges is

a) *Acacia* spp. b) *Erythrina* spp.

c) *Sesbania grantiflora* d) All

1292. Which is the type of functional basis for agro forestry classification?

a) Productive b) Protective

c) Aesthetic d) All

1293. Area under forest in India according to National Remote Sensing Agency (NRSA)

a) 75 mha b) 100 mha

c) 50 mha d) None

1294. Tree species suitable under coastal area is

a) *Casuarina equisetifoloa* b) *Cocus* spp.

c) *Borrassus flabellifer* d) All

1295. The ratio of height & width of shelter belt is

a) 1 : 10 b) 1 : 5

c) 1 : 3 d) 10 : 20

1296. Ideal character of tree species for alley cropping is

a) Fixation of nitrogen

b) Tolerant under abnormal conditions

c) Deep rooted

d) All

1297. Area under forest land in India is

a) 67 mha
b) 90 mha
c) 100 mha
d) 143 mha

1298. What is the major cause of deforestation

a) Shifting cultivation
b) Cutting trees for fuel wood
c) Both
d) None

1299. The term “organic” refers to

a) Living matrial
b) Non-living material
c) Mixture of both
d) None

1300. Most suitable plant for vegetative barrier

a) Merker grass
b) Khus grass
c) Guatemala grass
d) Ulu grass

1301. The most serious sheet erosion occurs in

a) Red and Black soils
b) Red and Alluvial soils
c) Black and Alluvial soils
d) Red and Laterite soils

1302. Which of the following mechanical soil conservation measures is more suitable for fruit trees or other plantation crops on steep slopes?

a) Contour bunds
b) Half moon terraces
c) Graded bunds
d) Bench terraces

1303. Land degradation first starts with the

a) Reduction in vegetative cover
b) Exposing the soil surface to accelerated erosion
c) Reduction in soil organic matter and nutrient content
d) All the above

1304. The primary objectives of tillage is

a) Seedbed preparation
b) Provision of a good medium for plant roots
c) Water infiltration and retention
d) Erosion and weed control

1305. Conservation tillage can be defined as

a) A crop planting system that allows minimum disturbance of the soil to allow seeds to be so shown while ensuring maintenance of crop residues on the surface

b) A tillage operation is mainly done to prevent the weed population

c) A tillage operation is mainly done to prepare a fine seedbed

d) All the above

1306. What is important feature of soil erosion?

a) Erosion removes top soil

b) Reduces levels of soil organic matter

c) It contributes to the breakdown of soil structure

d) All the above

1307. Long-term soil erosion results in

a) Persistent and large gullies

b) Exposure of lighter colored subsoil at the surface

c) Poorer plant growth

d) All the above

1308. Wind erosion is calculated by

a) Ramser's formula b) Chepil equation

c) Chepil and Woodruff equation d) All

1309. In north-eastern hill regions of India, shifting cultivation is also called as

a) Jhum cultivation b) Jhuming cultivation

c) Both d) None

1310 Total number of land capability classes in the land capability classification

a) 6 b) 7

c) 8 d) 9

1311. Total number of classes in the group, land suitable for cultivation is

a) 8 b) 4

c) 6 d) 2

1312. In land capability classification scheme, the class's V, VI and VII are suitable for

a) Cultivation

b) Pasture and grazing

c) Wildlife and watershed management

d) All the above

1313. Rills with more than ________ cm depth are generally called as gullies.

a) 18 b) 20

c) 25 d) 30

1314. Of the following, which one is important with regards to soil erosion by water?

a) Intensity of rainfall b) Duration of rainfall

c) Amount of rainfall d) Frequency of rainfall

1315. If the velocity of water is doubled, its erosion power increases ________ times, and carrying capacity increases ________ times.

a) 3 and 16 b) 4 and16

c) 4 and 32 d) 4 and 64

1316. An agronomist is primarily concerned with what use of soil?

a) Construction

b) Acting as a filter for the hydrologic cycle

c) Supporting crop growth

d) Natural beauty

1317. What is no-till crop production?

a) Preparing a good seedbed for crops

b) Ploughing ground, but not discing

c) Planting the seeds in the previous year's crop residue

d) Discing ground prior to planting

1318. CSWCRTI was established in which year?

a) 1974 b) 1949

c) 1958 d) 1985

1319. First recorded soil conservation research in India started in which year?

a) 1923 b) 1935
c) 1958 d) 1974

1320. First soil conservation scheme called Dry farming Scheme started atwhich place in India?

a) Manjiri near Pune b) Dehradun
c) Barapani at Shillong d) None of the above

1321. Which of the following is a type of terrace?

a) Diversion b) Retention
c) Bench d) All

1322. Soil erosion is a

a) Three stage process b) Two stage process
c) Five stage process d) None

1323. The uniform removal of thin layers of soil from a more or less smooth slope, carried by the distributed (rather than concentrated) flow of runoff water over the soil surface is called

a) Sheet erosion b) Gully erosion
c) Rill erosion d) None

1324. Splash and sheet erosion are also known as

a) Inter-rill erosion b) Inter-rill erosion
c) Both d) None

1325. The scouring and transport of soil by a concentrated flow of water is called

a) Rill erosion b) Sheet erosion
c) Both d) None

1326. Saltation occurs when particle size is

a) 0.1-0.5 mm b) 0.5-1 mm
c) 2-5 mm d) 10-20 mm

1327. Different types of models includes

a) Mathematical models b) Statistical models
c) Simulation models d) All

1328. The erosive power ot rainfall, running water, or wind is called

a) Erosivity b) Erodibility
c) Susceptibility d) Run off

1329. The erosive power of rainfall depends upon

a) Rainfall intensity b) Duration
c) Both d) None

1330. The erosive power in the case of running water depends on

a) The velocity b) Turbulence
c) Both d) None

1331. An equation relating the height of the water-table between drains to depth of the drains and the distance between them,the hydraulic conductivity of the soil, and the rate of recharge by percolating water is called

a) Hooghoudt equation b) Darcy equation
c) Renould's equation d) None

1332. The volume of water that runs off during an irrigation thatexceeds the soil's infiltration capacity is called

a) Run off b) Erosion
c) Splash erosion d) Sheet erosion

1333. A section of the land surface that drains its surface-water excess to a specified point along a stream is called

a) Watershed b) Catchment
c) Both d) None

1334. Vegetative windbreaks are also called

a) Shelterbelt b) Tree belts
c) Watershed d) None

1335. Water erosion is a

a) 2 way process b) 1 way process
c) 3 way process d) None

1336. Which of the following soil properties influences erodability?

a) Infiltration capacity b) Structural stability
c) Both d) None

1337. Agro forestry interventions with adequate soil and moisture conservation methods can reduce the soil loss by

a) 30% b) 40%

c) 50% d) 70%

1338. Which of the following erosion can be easily removed by normal tillage operations?

a) Rill erosion b) Gully erosion

c) Sheet erosion d) Splash erosion

1339. Most common size of raindrop is

a) 2 mm b) 3 mm

c) 4 mm d) 7 mm

1340. The maximum impact of the raindrop will be at what anglewhen it hits the ground?

a) 30° b) 45°

c) 90° d) 180°

1341. Which of the following soil separates has more erodability?

a) Sand b) Silt

c) Clay d) Rocks

1342. Which of the following types of clay minerals will provide more stable aggregates?

a) Nonexpanding type of clay minerals

b) Expanding type of clay minerals

c) Both

d) None

1343. Which is the most impoartn index of soil erodibility

a) Silica content b) Sesquioxide content

c) Silica-Sesquioxide ratio d) All

1344. The major role of inter-cultivation tillage is

a) Improvement of soil texture b) Enhancement of soil aeration

c) Weed control d) Improvement of soil structure

1345. According to "Universal soils loss equation the factors on which ratio of erosion depend are

a) Rainfall, soil credibility and slope length

b) Rainfall, soil erodibility, slope length, slop gradient, crop management & erosion control practice

c) Rainfall, wind velocity, crop management

d) Soil credibility, soil type, crop management

1346. Water harvesting refers to

a) Collection of runoff water into micro or macro reservoirs andrecycling the store water for irrigation

b) Conservation of rain water in soil by preventing the runoff somat rain water is better utilized in dry land

c) A technique of better harvesting the ground water to supplement the canal irrigation

d) A method of preventing percolation loss of water by suitably applying moisture barrier substances like asphalt.

1347. Pure and free water has greater potential than soil water

a) True b) False

c) Sometimes true d) Don't known

1348. Infiltration of soils with hard pan can be increased by

a) Basin listing b) Mulching

c) Sub soiling d) None

1349. Soil crushing reduces

a) Permeability b) Infiltration

c) Run off d) All

1350. Poor water holding capacity of red soils is mainly due to

a) High infiltration b) Shallow departments

c) Coarse texture d) None

1351. The downward flow of water from the surface into the soil is

a) Percolation b) Infiltration

c) Permeability d) Hydraulic conductivity

1352. Vertical down-ward movement of water within the soil is called

a) Infiltration
b) Percolation
c) Seepage
d) Capillary

1353. Clay content is higher in soil

a) Red
b) Desert
c) Black
d) All

1354. Soil erosion can be avoided by

a) Maintaining a protective cover on the soil
b) Creating a barrier to the erosive agent
c) Modifying the landscape to control runoff amounts and rates
d) All

1355. Which type of erosion is the most highly visible?

a) Gully erosion
b) Wind erosion
c) Ephemeral gullies
d) Splash erosion

1356. The largest of those soil separates or particles

a) Gravel
b) Sand
c) Silt
d) Clay

1357. Which of these is not a key ingredient of soil?

a) Organic matter
b) Haemoglobin
c) *Minerals*
d) Air

1358. What percentage of the average soil is organic matter?

a) 45%
b) 5%
c) 25%
d) 17%

1359. Erosion created by the activities of man and sometimes by animals is called as

a) Natural geological erosion
b) Geologic erosion
c) Accelerated soil erosion
d) All

1360 Scattering of detached soil particles by the impact of rain drops is called

a) Splash erosion
b) Sheet erosion
c) Gully erosion
d) Ravines

1361. Sheet erosion is observed on

a) Deeply sloping land b) Moderate sloping land
c) Sloppy land d) Gently Sloppy land

1362. Muddy run-off from the field is an indication of

a) Splash erosion b) Sheet erosion
c) Rill erosion d) Ravines

1363. Formation of small channels a few centimeters deep al over the field along the watercourse is observed in

a) Rill erosion b) Slightly Gully erosion
c) Moderately Gully erosion d) Ravine

1364. Erosion permitting crops is

a) Maize b) Wheat
c) Bajra d) All

1365. Erosion resisting crops

a) Grasses b) Legumes
c) Cereals d) Both a and b

1366. Removal by rain of a very thin layer of soil from the entire surface of large area is called

a) Splash erosion b) Sheet erosion
c) Rill erosion d) Ravines

1367. The ratio of dispersion ratio to colloidal moisture equivalent ratio is

a) Erosion index b) Erosion ratio
c) Soil detachability d) Soil erosivity

1368. In Universal soil loss equation, A=RKLSCP, K denotes

a) Soil conservation practices b) Soil roughness
c) Soil erodibility d) Soil erosivity

1369. In Universal soil loss equation, A=RKLSCP, P denotes

a) Soil conservation practices b) Soil roughness
c) Soil erodibility d) Rainfall factor

1370. Growing of erosion resisting and erosion permitting crops on alternate strips of suitable width along the contour and across the slope is called as

a) Contour cropping b) Contour farming
c) Contour strip cropping d) Strip cropping

1371. Contour bunding is also called as

a) Level terraces b) Ridge-type terraces

c) Absorption type terraces d) All

1372. Soils more susceptible to water erosion are

a) Course textured soils b) Medium textured soil

c) Fine textured soils d) All

1373. Minimum wind speed required to initiate the movement of most erodible soil particle is

a) 10 km b) 12 km

c) 14 km d) 16 km

1374. Diameter of particles that is most susceptible to wind erosion is

a) 0.1 mm b) 0.2 mm

c) 0.02 mm d) 0.01 mm

1375. Dominant type of wind erosion in soil is

a) Saltation b) Surface creep

c) Suspension d) All

1376. Size range of soil particles in saltation type of wind erosion is

a) <0.05 mm b) 0.5-1 mm

c) 0.5-2 mm d) 0.05-0.5 mm

1377. Size range of soil particles in suspension type of wind erosion is

a) <0.05 mm b) 0.5-1 mm

c) 0.5-2 mm d) 0.05-0.5 mm

1378. Size range of soil particles in surface creep type of wind erosion is

a) <0.05 mm b) 0.5-1 mm

c) 0.5-2 mm d) 0.05-0.5 mm

1379. In surface creep, soil particles are moved primarily by the action of

a) Wind b) Saltation

c) Suspension d) All

1380. Most important factors that influence the erodability of soil by wind

a) Soil texture b) Soil structure

c) Soil water status d) All

1381. The crops which help to effectively check soil erosion and conserve soil moisture and nutrients are referred to as

a) Cover crops
b) Buffer crops
c) Erosion crops
d) Conservative crops

1382. Cover crops help to check soil erosion and conserve soil moisture and nutrients by

a) Checking movement of run-off water
b) Increasing absorption of rain water by soil
c) Soil cover is required against direct hitting of raindrops and protection against the velocity of wind
d) All these ways

1383. Which of the following crops are most effective in protecting cultivated land from erosion

a) Legumes
b) Millets
c) Cereals
d) All

1384. The type of crop and cropping system that can be used to conserve soil against erosion varies from region to region depending on

a) Soil type
b) Climate
c) Both
d) None

1385. Which of the following crops are known to provide effective cover against loss of soil nutrient and moisture

a) Cowpea, mung, urad
b) Dhaincha, sunhemp, groundnut
c) Velvet bean, guar, soyabean
d) All

1386. At Kanpur, the best vegetative cover against soil and water loss was provided by

a) Velvet bean
b) Cowpea
c) Mung
d) Soybean

1387. Water harvesting is defined as

a) Application of water at the time of harvesting cropslike colocasia, tapioca and yam
b) Harvesting deep water from soils by pumping devices
c) Unit of water applied to crops
d) Collecting the excess run off from rain on the farm in ponds and Utilizing it later for agriculture

1388. At Sholapur in Maharashtra the most effective check against soil and water loss was provided by

a) Moth bean b) Horse gram

c) Both a & b d) Cowpea

1389. Which of the following conditions would lead to the most soil erosion?

a) Heavy rainfall (from a single storm) in an area of low average annual rainfall.

b) Steep slopes in a tropical rainforest

c) An agricultural region where soil conservation practices are implemented

d) An area that has been urbanized for several decades or more.

1390. Why do agricultural areas often experience the greatest amount of soil erosion?

a) Because they typically occur in areas of high rainfall.

b) Because they typically occur in areas of moderately steep slopes.

c) Because farming typically involves plowing the soil to plant seeds.

d) Because farmers typically remove trees from their fields.

1391. What does the term "sediment yield" refer to?

a) The amount of sediment eroded from hilt slopes by water.

b) The amount of sediment eroded from the land by wind

c) The amount of sediment eroded by either water or wind

d) The amount of sediment leaving a drainage basin in a stream

1392. Why is sediment yield often less than soil erosion in a given drainage basin?

a) Because streams usually cannot carry all of the sediment that is eroded.

b) Because sediment yield is totally unrelated to soil erosion.

c) Because streams naturally evolve towards a graded condition.

d) Because most drainage basins occur in areas of high rainfall.

1393. Why were soil conservation efforts in Coon Creek Valley, Wisconsin notsuccessful in reducing Coon Creek's sediment yield?

a) Because the soil conservation efforts were ineffective

b) Because Coon Creek adjusted to the reduced soil erosion by erodingsediments from the valley bottom

c) Because the soil conservation efforts were not implemented throughoutthe entire drainage basin.

d) Because sediment yield is not related to soil erosion.

1394. Why might a river's suspended-sediment yield not be a good measure of thetotal soil erosion in that river's drainage basin?

a) Because some suspended sediment might come from other sources.

b) Because the river might not be able to carry all of the soil that is eroded.

c) Because the river might not be graded at the time when the suspended-sediment yield measurements are made.

d) All

1395. Under what conditions could the addition of weight to a slope trigger a *block glide?*

a) When the slope is composed of unconsolidated material.

b) When there is cohesion present along the potential slip plane

c) When the potential slip plane is oriented away from the valley

d) When the slope has just experienced a fire

1396. Which of the following slope movements could be triggered by heavy rainfall or snowmelt?

a) A slump

b) A block glide

c) A mudflow

d) All

1397. How does a *debris flow* differ from soil erosion by water?

a) They differ primarily by the amount of water involved in the process

b) Soil is not considered "debris".

c) Debris flows only occur on ash-covered volcanic slopes.

d) Actually, there is no difference between these processes

1398. Soil creep:

a) Occurs only on moderate to steep slopes.

b) is important mostly because of the damage it causes to human structures

c) creates "tracks" or "chutes" along hill slopes

d) is most effective in desert environments

1399. Debris flows:

a) can be triggered by periods snowmelt in the spring

b) typically have a *threshold* relationship with rainfall duration

c) often occur after forest fires move through mountainous areas along slopes that have little or no vegetation

d) All

1400. Which of the following slope movements would be most likely to occur as a result of adding weight to the "head" of the slope?

a) A lump b) A block guide

c) A soil creep d) A mud flow

1401. Concentrate organic manure

a) Mustard cake b) FYM

c) Vermicompost d) Phosphocompost

1402. FYM contains _________ N (%)

a) 0.5 b) 1.0

c) 0.75 d) 1.2

1403. IRRI is located in

a) Indonessia b) Malayasia

c) Phillippines d) Thiland

1404. Incorporation of green manure crops in soil is generally done ________ DAS.

a) 30 b) 45

c) 50 d) 60

1405. Mustard cake contains N (%)

a) 5 b) 6

c) 6.5 d) 4.5

1406. The optimum harvesting time of babycorn is

a) 1-2 days before fertilization b) 1-2 days after fertilization

c) Milk stage d) Early dough stage

1407. *Trichoderma is a fungi which decompose*

a) Cellulose b) Lignin

c) Hemicellulose d) Protein

1408. Optimum spacing of rice under SRI is

a) 25 cm x 25 cm b) 20 cm x 15 cm

c) 30 cm x 25 cm d) 25 cm x 15 cm

1409. Emergence of female flower in maize is called

a) Tasselling b) Silking

c) Blooming d) Flowering

1410 Basic slag is a suitable phosphatic fertilizer for

a) acid b) alkine
c) saline d) saline alkali soil

1411. Suitable bee species for apiculture

a) *Apis melifera* b) *Apis cerana indica*
c) *Apis dorsata* d) All

1412. Dual purpose cattle breed

a) Haryana b) Sahiwal
c) Red sindhi d) Gir

1413. *Bombyx mori* is silkworm of

a) Tasar silk b) Muga silk
b) Mulberry silk c) Eri silk

1414. Father of Modern Organic Agriculture is

a) Albert Howard b) Nicholas Lampkin
c) Lord Northbourne d) Masanobu Fukuoka

1415. Buffer zone is maintained in certified organic farming for

a) contamination control b) pest control
c) pollination control d) grazing animal control

1416. Laser land levelling is an example of

a) precision farming b) conservation agriculture
c) ecological farming d) LEISA farming

1417. In a biogas plant, the predominant gas generated is

a) methene b) ethelene
c) aceteline d) CO_2

1418. Judicious thinning at 20 -25 days after germination of sunflower is needed to maintain

a) single healthy plant per hill b) increased size of head
c) less disease incidence d) less incidence of lodging

1419. Water requirement of sugarcane for optimum growth varies from

a) 400-600 cm b) 200-300 cm
c) 600-700 cm d) 450-550 cm

1420. Stem nodulating green manure crop is

a) *Sesbania aculeata* b) *Sesbania rostrata*

c) *Crotolaria juncea* d) All

1421. Concentrate organic manure

a) Bone meal b) FYM

c) vermicompost d) All

1422. Jhum cultivation is called

a) slash and burn cultivation b) jabo cultivation

c) natural farming d) ecological farming

1423. Polyculture is associate with

a) mixed fish culture b) mixed vegetable cultivation

c) crop livestock mixed culture d) crop vegetable mixed culture

1424. The symbiosis phenomena is present in

a) Azolla b) BGA

c) *Azotobactor* d) *Azospirillum*

1425. Summer cultivation is a .... pest control.

a) cultural b) mechanical

c) biological d) chemical

1426. More than _____ per cent of the earth's surface are covered with water.

a) 55 b) 70

c) 76 d) 80

1427. In semi-arid regions, availability of rainfall / annum is ______ mm.

a) 200-400 b) 250-500

c) 300-550 d) 600-700

1428. On an average, ________days is wet days in a year in our country (India).

a) 130 b) 145

c) 155 d) None

1429. _________________ irrigation method is best for undulated topography

a) Check basin b) Flow irrigation

c) Sprinkler d) Furrow irrigation

1430. Root depth of potato considered for irrigation is

a) 15 cm b) 30 cm
c) 45 cm d) 60 cm

1431. Green fodder of oat contains_______% protein.

a) 5-7 b) 15-20
c) 10-12 d) 20-23

1432. Seed rate of Lucerne is ____________ kg/ha.

a) 10-12 b) 15-18
c) 20-24 d) 25-28

1433. The average green forage yield of Berseem is_____________ t/ha.

a) 80-100 b) 110-150
c) 150-180 d) 160-200

1434. Which of the following is a variety of Lucerne?

a) Anand-2 b) OS-6
c) Jho-822 d) P 3396

1435. Usual sowing time of lentil in eastern part of India is –

a) Mid Oct. to Mid Nov.
b) 3rd week of Nov. to 1st week of Dec.
c) 2nd fortnight of Dec.
d) None

1436. Layout of crop field is _________.

a) Primary tillage b) Inter cultivation
c) Preparatory tillage d) All

1437. Soil physical property affected by tillage operation

a) Bulk density b) Soil moisture
c) Soil colour d) All

1438. Stickiness or flakiness of cooked rice is largely determined by____ content

a) Protein b) Amylase
c) Vitamin d) Fat

1439. __________encourage natural cycle in organic farming.

a) Cereals b) Millets
c) Legumes d) None

1440. Inclusion of *Stylosanthes hamata* as a pasture legume in rotation with castor or sorghum is referred as__________________

a) Intensive cropping b) Sequence cropping
c) Relay cropping d) Ley farming

1441. Non- Polar herbicides are

a) More soluble in water b) Less soluble in water
c) Not soluble in water d) None of these

1442. Herbicide selectivity is a phenomenon of

a) Killing target plants only without or less affecting crop plants
b) Safe to crop plants
c) Both
d) None

1443. *Bhagirathi* is a variety of

a) Toria b) Rapeseed c)
Mustard d) Black mustard

1444. For weed management in zero tillage _________ is generally used

a) Glyphosate 71 SG b) Paraquat dichloride 24 SL
c) Diuron d) None

1445. The important nutrients for oilseed crops are

a) N & P b) N, P, K & S
c) N, P & K d) P & K

1446. *Rhizobium* inoculation is essential for following oilseed crops

a) Niger & Castor b) Rapeseed – Mustard
c) Sunflower & Safflower d) Soybean & Groundnut

1447. Dyes are made from the following plant part of safflower

a) Tender shoots b) Seeds
c) Flowers d) Roots

1448. The entire process from rough rice to head rice is considered to determine ____________ quality.

a) Milling b) Eating
c) Nutritional d) Processing

1449. Most common agro-forestry system in sub-tropical region is

a) Agri-silviculture
b) Silvipasture
c) Horti-silviculture
d) Agro-horti silviculture

1450. Maize-Wheat-Cotton is an example of

a) Mixed cropping
b) Inter cropping
c) Multiple cropping sequence
d) All

1451. Mixed cropping of Arhar with Jowar reduces the disease

a) Root rot of Jowar
b) Leaf blight of Arhar
c) Downy mildew of Arhar
d) Wilt of Arhar

1452. Plant quarantine is practical application of Principle

a) Eradication
b) Therapy
c) Exclusion
d) Avoidance

1453. Presence of diseased plants in the infected field or orchard is a continuous source of

a) Insect
b) Nematode
c) Fungus
d) Infective propagule

1454. Extensive mono cropping of susceptible crop in an area for long time results in

a) Destruction of collateral host
b) Easy survival of the pathogen
c) Decomposition of the organic matter
d) Enrichment of nutrition in soil

1455. Decomposition of organic matter through organic amendment alters

a) Physical environment of soil
b) Chemical environment of soil
c) Physical,chemical and biotic environment of soil
d) None

1456. The disease spread fast in close plant spacing is

a) Wilt
b) Root rot
c) Virus
d) Leaf spot

1457. The resting structure produced by fungal pathogen survives in soil are

a) Ascospore b) Chlamydospore

c) Sporangiospore d) Basidiospore

1458. Disease escaping quality of a variety is due to its

a) Genetic character

b) Characteristics of growth and time of maturity

c) Susceptibility towards the pathogen

d) Environmental influence

1459. Biological control practices fall under the principle

a) Resistance b) Exclusion

c) Therapy d) Eradication

1460 Bunchy top disease of banana is an endemic disease in the state

a) TN b) Kerala

c) AP d) Karnataka

1461. After enforcement of DIP act in India during 1914, the disease introduced was

a) Late blight of potato b) Downy mildew of grapes

c) Wheat rust d) BLB of rice

1462. Flood fallowing method of disease management is mainly adopted in

a) Damping off b) Root rot

c) Wilt of banana d) Blight

1463. Many generations of pathogen cause infection in one life cycle of crop is called

a) Pathogenesis b) Heterokaryosis

c) Monocyclic disease d) Polycyclic disease

1464. The disease constantly present in some locality for long time is called

a) Pandemic b) Sporadic

c) Epidemic d) Endemic

1465. The rouging practice can be adopted to check the spread of the disease

a) Blight b) Root rot

c) Loose smut d) Damping off

1466. Through decomposition of Brassica plants the toxic chemical released is

a) Ammonia b) HCN

c) Phenol d) Glucosinolate

1467. The fungicide withdrawn from the market due to undesirable toxicological problem is

a) Antracol b) Fytolan

c) Metalaxyl d) Edifenphos

1468. An ideal and widely recommended fungicide to control blast disease of paddy is

a) Mancozeb b) Bavistin

c) Blitox d) Tricyclazole

1469. The earliest plant quarantine law was established in

a) India b) Germany

c) USA d) France

1470. Potato-Garlic mixed cropping helps to manage the disease

a) Wilt of potato b) Common scab of potato

c) Black scurf of potato d) Root knot disease of potato

1471. Crop rotation forces the plant pathogens to

a) Multiply b) Transmit

c) Infect the host d) Persist as survival structure

1472. The fungicide PCNB is now banned but earlier it was extensively used to manage the

a) Seed borne pathogen b) Wilt causing pathogen

c) Blight causing pathogen d) Soil borne pathogen

1473. Through execution of International Plant Qurantine laws, the karnal bunt disease of wheat is restricted in

a) Pakistan b) Thailand

c) Japan d) India

1474. The study of factors affecting the outbreak and spread of infectious diseases is

a) Ecology b) Etiology

c) Mycology d) Epidemiology

1475. The crop variety showing low level of economic loss due to their high yielding capacity even when infected is called

a) Resistant
b) Susceptible
c) Immune
d) Tolerant

1476. Heteroecious nature of rust pathogen was first recognized by

a) L.M.Joshi
b) K.C.Mehta
c) E.J.Butler
d) Anton de Bary

1477. *Puccinia graminis tritici* overwinters as

a) Basidiospore
b) Pycniospore
c) Aeciospore
d) Teliospore

1478. Micrometer (μ) is a unit of length equal to

a) 1/10 of a millimeter
b) 1/100 of a centimeter
c) 1/10 of a meter
d) 1/1000 of a millimeter

1479. The sheath blight of paddy is effectively managed by foliar spray of a bioantagonist i.e,

a) *Trichoderma harzianum*
b) *Bacillus cereus*
c) *Gliocladium virens*
d) *Pseudomonas fluorescens*

1480. The kresek symptom is produced in

a) Rice tungro virus disease of paddy
b) BLB disease of paddy
c) Blast disease of paddy
d) Sheath blight of paddy

1481. Degenerated growth is a common symptom in case of disease

a) Downy mildew of Jowar
b) Rice tungro virus of paddy
c) Wheat rust
d) Leaf blight of maize

1482. For a BLB pathogen a weed host considered as most important active natural host is

a) *Cyperus* sp.
b) *Brachiaria* sp.
c) *Panicum* sp.
d) *Leersia* sp.

1483. Scientist Luthra is famous for devising a low cost management strategy for the disease

a) Maize stalk rot
b) Brown spot of rice
c) Wheat rust
d) Loose smut of wheat

1484. Dikaryotic mycelium is produced during the life cycle of the pathogen
a) *Phytophthora* sp b) *Helminthosporium* sp
c) *Puccinia* sp d) *Alternaria* sp

1485. Validamycin is nowadays has been prescribed by the plant pathologist for managing the disease
a) False smut of paddy b) Wheat rust
c) Barley yellow dwarf d) Sheath blight of paddy

1486. Now-a-days, a routine diagnostic approach to assay the starch accumulation in the infected leaves is studied for detection of the pathogen causing
a) Rice tungro virus disease b) Maize leaf blight
c) Wheat rust d) Sorghum leaf spot

1487. *Xanthomonas oryzae* takes entry within the host through
a) Stubbles b) Hydathodes
c) Straw d) Weed

1488. Through management of irrigation practices the effective check in further spread is possible for the disease
a) Rice tungro virus disease b) Sheath blight of paddy
c) Brown spot of paddy d) BLB of paddy

1489. In case of Black stem rust disease of wheat, infection of alternate host is done by
a) Teliospore b) Pycniospore
c) Basidiospore d) Aeciospore

1490. PR-3 rice chitinase gene has been successfully incorporated into the susceptible cultivar to develop the resistance against
a) Wheat rust pathogen b) Lose smut pathogen
c) BLB pathogen d) Sheath blight of paddy

1491. *Rhizoctonia solani* is not a single species but is composed of groups based on
a) Pathogenecity b) Reproduction
c) Anastomosis d) Vegetative growth

1492. Bacterial stalk rot disease of maize is caused by
a) *Pseudomonas* sp. b) *Xanthomonas* sp.
c) *Bacillus* sp. d) *Erwinia* sp.

1493. Organic amendment of soil can manage the plant pathogen

a) *Alternaria solani* b) *Helminthosporium oryzae*

c) *Magnaporthe grisea* d) *Rhizoctonia solani*

1494. Flood irrigation helps to suppress the survival structures of the plant pathogen

a) *Fusarium* sp. b) *Cercospora* sp.

c) *Ustilaginoidea* sp. d) *Helminthosporium* sp.

1495. Plant disease management through biological control practices fall under the principle

a) Aviodance b) Exclusion

c) Eradication d) Resistance

1496. Length of crop rotation for managing a soil borne plant pathogen depends on

a) Pathogenecity of the pathogen

b) Age of the pathogen

c) Period of perpetuation of the resting structures of the pathogen

d) Colour of the pathogen

1497. A systemic fungicide is

a) Chlorothalonil b) Hexaconazole

c) Antracol d) Thiram

1498. A combined formulation product of fungicide is

a) Bavistin b) Metalaxyl

c) Calyxin d) Curzate

1499. In case of Black stem rust of wheat disease on alternate host the spore forms produced are

a) Teliospore & aeciospores b) Basidiospore & uredospore

c) Aeciospores & pycniospore d) Pycniospore & uredospore

1500. In case of wheat rust disease, n+n containing teliospore is converted to containing basidiospore through

a) Plasmogamy b) Karyogamy

c) Meiosis d) Karyogamy followed by meiosis

1501. Removal of diseased plants from the field is known as

a) Prunning b) Surgery

c) Roughing d) Training

1502. Crop rotation is the most effective methods for controlling

a) Root diseases b) Virus diseases

c) Foliar diseases d) Fruit diseases

1503. Mancozeb is a

a) Systemic fungicide b) Contact fungicide

c) Copper fungicide d) None of these

1504. Roguing is employed for controlling

a) Loose smut of wheat b) Rust of wheat

c) Blast of rice d) Leaf blight of wheat

1505. Flooding of field is done for controlling

a) Sigatoka of banana b) Bunchy top of banana

c) Moko disease of banana d) Panana wilt of banana

1506. Wilt (*Fusarium*) of pigeon pea can be chacked by treating the seed with

a) Thiram b) Ceresan

c) Blitox d) None of these

1507. Yellow mosaic virus disease of mungbean is transmitted in the field by

a) Leaf hopper b) Whitefly

c) Aphid d) Thrips

1508. Flag smut of wheat can be successfully checked by adopting crop rotation with non susceptible crops like

a) Barley b) Maize

c) Rice d) None of these

1509. Seed treatment of wheat before sowing with carbendazim provide protection to

a) Loose smut of wheat. b) Alternaria leaf blight of wheat

c) Karnel bunt of wheat d) None of these

1510 Red rot of fungus (Colletotrichum falcatum) can persist in soil for

a) Few months b) 1 year

c) 2 years d) 3 years

1511. Biochemical test for diagnosis of bacterial wilt of potato is done with the help of chemical namely,

a) NaCl
b) NH_4 Cl
c) KOH
d) NH_4 NO_3

1512. In case of standing crop, if bacterial wilt infection is noticed, then immediate measure to be adopted to check the further spread of the disease in the field is

a) Foliar application of systemic fungicide
b) Foliar application of contact fungicide
c) Roguing of infected plant
d) Application of irrigation water

1513. Strobilurin being novel group of fungicide is developed from a mushroom namely

a) *Pleurotus* sp.
b) *Volvariella* sp.
c) *Agaricus* sp.
d) *Strobilurus* sp.

1514. The supreme authority to ban or withdraw a fungicide from the market is

a) Plant quarantine department
b) Central Insecticide Board
c) Ministry of Agriculture, Govt. of India
d) State Agriculture University

1515. Roguing is an important cultural practice to manage the disease i.e.

a) Sheath blight of paddy
b) Potato late blight
c) Loose smut of wheat
d) Wheat leaf blight

1516. The fungicide acting as melanin biosynthesis inhibitor is

a) Calyxin
b) Tricyclazole
c) Myclobutanil
d) Bavistin

1517. Most common fungicide earlier used as soil treatment but nowadays banned is

a) Mancozeb
b) Captan
c) Brassicol
d) Hinosan

1518. Mode of action of triazole group of fungicide is to inhibit

a) Cell function
b) Nucleic acid synthesis
c) Sterol synthesis
d) Protein synthesis

1519. First generation fungicide developed during represent

a) Carboxin b) Edifenphos

c) Copper based fungicide d) Thiophanate methyl

1520. The sheath blight of paddy is effectively managed by foliar spray of bio antagonist i.e.

a) Trichoderma harzianum b) Bacillus cereus

c) Pseudomonas fluorescens d) Gliocladium virens

1521. Organic amendment of soil can manage the plant pathogen i.e,

a) Alternaria solani b) Helminthosporium oryzae

c) Rhizoctonia solani d) Cercospora solani

1522. A systemic fungicide is

a) Antracol b) Hexaconazole

c) Chlorothalonil d) Blitox

1523. Length of crop rotation for managing a soil borne pathogen depends on

a) Pathogenecity of the pathogen

b) Capacity of the pathogen to infect the host

c) Type of spore produced by the pathogen

d) Duration of survibility of the resting structures of the pathogen

1524. A widely used seed treating contact fungicide is

a) Bordeaux mixture b) Propiconazole

c) Thiram d) Calyxin

1525. In crop rotation practices there must be incorporation of

a) Wild cultivar b) Collateral host

c) Resistant or non host d) Susceptible host

1526. High ridge planting reduces the ________ disease

a) Powdery mildew b) Bacterial wilt

c) Virus d) Rust

1527. Early planting of potato helps to escape the – disease

a) Scab of potato b) Early blight of potato

c) Black scurf of potato d) Late blight of potato

1528. Bleaching powder can be applied with irrigation water to check the further Spread of the disease i.e.

a) Sheath rot

b) Leaf blight of wheat

c) Root k not nematode disease of vegetables

d) Bacterial wilt of potato

1529. Through eradication of Alternate host barbery, the wheat stem rust pathogen

a) Can easily infect the wheat host

b) Can not complete its life cycle

c) Can form all the spores in the main host

d) Can produce large number of inoculums

1530. Use of fresh cow dung slurry after sieving it in fine cloth and sprayed 3-4 times at

a) Sheath rot of paddy

b) Leaf blight of wheat

c) Bacterial leaf blight of paddy

d) Brown spot disease of paddy

1531. Dew deposition is an important environmental factor that significantly favors the disease

a) Sheath blight of paddy

b) Blast of paddy

c) Rice tungro virus disease of paddy

d) Stem rot of paddy

1532. Presence of diseased plants in the field or orchard is a continuous source of

a) Insects

b) Fungus

c) Nwmatodes

d) Inoculam

1533. Extensive mono cropping of susceptible host in an area results in

a) Destruction of weeds

b) Decomposition of organic matter

c) Enrichment of nutrition in the soil

d) Easy survival of the pathogen

1534. Mixed cropping of Arhar with Jowar reduces the disease i.e,

a) Leaf blight of Arhar
b) Downy mildew of Arhar
c) Wilt of Arhar
d) Root rot of Arhar

1535. An antibiotic nowadays widely used for to manage the Sheath blight disease of Paddy is

a) Streptomycin
b) Penicillin
c) Blasticidin
d) Validamycin

1536. Biological protection against infection is accomplished by

a) Destroying the existing inoculum only
b) Preventing the formation of additional inoculums
c) Weakening and displacement of the existing virulent pathogen population.
d) All

1537. dsRNA introduction in the fungal pathogen population could effectively reduce the ________ disease

a) Root rot of pea
b) Chestnut blight
c) Sheath rot of rice
d) Stem rot of chickpea

1538. Biocontrol fungus *Trichoderma* spp could effectively be isolated from

a) Submerged paddy rhizosphere
b) Healthy plant rhizosphere soil in heavily infected field
c) Phyllone
d) From all these sources

1539. Biocontrol agents concepts for nutrients on phyloplane. Which nutrient element is most important?

a) P
b) Fe
c) K
d) N

1540. Yeast is better candidate than any filamentous biocontrol fungus to control diseases on phyloplane because

a) It require less amount of nutrient
b) It can tolerate radiation damage
c) Both
d) None

1541. Enhancement of resident soil and plant associated biocontrol microorganisms could be possible by

a) Selective elimination of resident soil borne plant pathogens by soil solarization

b) Using suppressive compost

c) Both

d) None

1542. For selective isolation of *Trichoderma* spp from soil by using a medium which should contain the fungicide

a) Carbendazim
b) Captan
c) Carboxin
d) Mancozeb

1543. PGPR protects plants from diseases by

a) Lytic enzyme production
b) Siderophore production
c) Antibiosis
d) Induced resistance

1544. Biocontrol fungus *Trichoderma* spp produce spores in culture media under the condition

a) In rich medium and under dark condition

b) Low nutrient containing media & dark condition

c) At $28^0 \pm 2^0$C with brief exposure to light

d) At 35-40°C

1545. In dual culture plate with a fungal plant pathogen and a biocontrol agent, it was observed that due to presence of biocontrol agent growth of the pathogen was ceased with a sharp margin towards the bioagents but far from the biogent growth. What would be the possible reason?

a) Mycoparasitism
b) Depletion of nutrient
c) Siderophore production
d) Antibiosis

1546. For getting good result in plant disease biocontrol, which one of the following is the best?

a) Use of the native bioagent

b) Introduction of foreign isolate with high potential

c) Use of bioagent consortium with native isolate

d) Use of a highly potential bioagent once in a year

1547. Among the fungal biocontrol agents *TrIchoderma* Spp are exploited most. Which of the following reason is most appropriate

a) Widely available

b) Easily isolated and multiplied

c) Widely available and easily multiplied

d) Easily recognizable

1548. *Trchoderma* spp is an amarmorphic fungus. The teleomorphic phase of them are belong to the fungal class

a) Oomycetes b) Ascomycetes

c) Zygomycetes d) Basidiomycetes

1549. *Trchoderma* spp could be recognized easily on culture medium MEA due to their

a) Slow growing, greenish colony, spores produced uniformly on the medium

b) Fast growing with green sporulation in zone

c) White, cottony with yellow pigmentation in the medium

d) Cottony mass with black sporulation

1550. Mycorrhizal fungi provide protection to the plants against plant diseases mostly by

a) Competition for nutrient b) Competition for space

c) Modifying the rhizosphare. d) All the above

1551. Which of the following is TPS variety of Potato?

a) JH 222 b) Chipsona-II

c) Anand d) HPS-1/113

1552. VAM are symbiotic ________ that infect plant roots

a) Viruses b) Bacteria

c) Algae d) Fungi

1553. What is the main function of zinc in the plants?

a) Synthesis of nitrogen

b) Synthesis of phosphorus

c) Required for synthesis of Tryptophos

d) To increase activity of the boron

1554. Which is the correct sequence of soil erosion?

a) Rill – Sheet – Gulley
b) Gulley – Sheet – Rill
c) Sheet – Rill – Gulley
d) Sheet – Gulley – Rill

1555. Zinc Sulphate ($ZnSO_4$) should not be mixed with

a) DAP
b) Compost fertilizer
c) Ammonium Chloride
d) Urea

1556. Insecticides are specific inhibitors of

a) Excretory system
b) Digestive system
c) Nervous system
d) Blood Circulatory system

1557. The credit for the success of Krishi Vigyan Kendras (KVK) goes to—

a) Dr. R. S. Paroda
b) Dr. M.S. Swaminathan
c) Dr. Mohan Singh Mehta
d) Dr. Mangla Rai

1558. Which of the following factors does not affect the nitrification?

a) Air
b) Seed
c) Temp
d) Moisture

1559. The permanent preservative, which is used for preservation of fruit and vegetables, is ________

a) Sodium chloride
b) Potassium metabisulphate
c) Potassium sulphate
d) Sugar

1560. *Motha* (Grass nut) belongs to the family of ________

a) *Cruciferae Brassicaceae*
b) *Tiliaceae*
c) *Cyperaceae*
d) *Poaceae*

1561. Which of the followings are short day crops?

a) Maize, Lobia, Bajra
b) Wheat, Mustard, Gram
c) Mung, Soybean, Bajra
d) Wheat, Soybean, Bajra

1562. What is the sequence of C_4 plants?

a) Sudangrass – Sugarcane –Paddy – Bajra
b) Sugarcane – Maize – Sudangrass – Bajra
c) Sugarcane – Cotton – Paddy – Maize
d) Cotton – Maize – Bajra –Sugarcane

1563. In which state, are there biggest area, highest production and number of Sugar Mills in relation to Sugarcane?

a) Maharashtra b) Bihar
c) UP d) AP

1564. Which is not prepared by potato?

a) Acetic Acid b) Paper
c) Wine d) Famina

1565. India is the fourth largest cotton producing country of the world after

a) Russia, China, USA b) USA, China, Russia
c) China, USA, Russia d) USA, Russia, China

1566. Growing different crops one after the other on the same field during different crop seasons is called

a) Mixed farming b) Strip cropping
c) Inter cropping d) Multiple cropping

1567. Which of the following countries have very small agricultural land holding?

a) France b) Japan
c) China d) India

1568. The factors and geographical areas responsible for India to depend on agriculture are

a) Vast cultivable land b) Wide climatic range
c) Long growing season d) All the above

1569. Millets are called 'coarse grains' and constitute mainly of

a) Maize, Jowar, Pulses b) Jowar, Bajra, Ragi
c) Bajra, Ragi, Maize d) Maize, Jowar, Ragi

1570. Research Institute on dairy development is at

a) Amritsar b) Karnal
c) Lucknow d) Kanpur

1571. Which of the following statements is incorrect?

a) The size of land holdings is small in intensive type of cultivation.

b) Mechanized farming is practiced in shifting agriculture.

c) Farming is done on a large scale resembling that of the factory production is plantation agriculture

d) All

1572. How much cultivated area has been brought under irrigation?
   a) 20% b) 25%
   c) 26% d) 28%

1573. What is the major factor responsible for degradation of environment?
   a) Industrialization b) Urbanization
   c) Both a & b d) None

1574. India's main rival in tea export is
   a) Japan b) Sri Lanka
   c) China d) USA

1575. Tilt angle of a disc plough is generally ________
   a) 10° b) 15°
   c) 20° d) 45°

1576. Pudding is done to ________
   a) Reduce percolation of water b) Pulverise and levelling soil
   c) Kill weeds d) All

1577. The Community Development Programme (CDP) was started in India on ________
   a) 2nd October, 1950 b) 2nd October, 1952
   c) 2nd October, 1951 d) None

1578. The main unit of Integrated Rural Development Programme is ________
   a) Family b) Village
   c) Block d) District

1579. Element of Communication is ________
   a) Message b) Feedback
   c) TV Channel d) All of these

1580. The main function of NABARD is ________
   a) Farmers' loaning b) Agricultural research
   c) Refinancing to agricultural financing institutions d) Development of agriculture

1581. Rent theory of profit was given by ________
   a) Hawley b) C.P. Blacker
   c) Tanssig d) F.A. Walker

1582. Acid rain contains mainly ________

a) PO_4 b) NO_2

c) NO_3 d) CH4

1583. Endosulphan is also known as ________

a) Lindane b) Thiodan

c) Aldrin d) B.H.C.

1584. Which of the following is systemic poison?

a) Metasystox b) Phosphomidan

c) Phorate d) All

1585. DDVP is known as ________

a) Nuvan b) Malathion

c) Thiodan d) Sulfex

1586. Seed treatment with Vitavex is the main controlling method of _______

a) Loose smut b) Rust

c) Downy mildew d) All

1587. Covered smut of barley is a disease of ________

a) Externally seed-borne b) Internally seed-borne

c) Air-borne d) None of these

1588. Which of the following cakes is not edible?

a) Castor cake b) Mustard cake

c) Sesame cake d) Groundnut cake

1589. In India, about 142 million hectare land is under ________

a) Cultivation b) Waste land

c) Forest d) Eroded land

1590. The headquarters of Indian Meteorological Department was established in 1875 at

a) New Delhi b) Hyderabad

c) Pune d) Calcutta

1591. Moisture condensed in small drops upon cool surface is called _______

a) Hail b) Dew

c) Snow d) Fog

1592. How many agro-climatic zones (ACZ) are found in W.B.?

a) 5 b) 6

c) 7 d) 8

1593. 'Morden' is a mutant variety of

a) Wheat b) Sunflower

c) Safflower d) Cotton

1594. In sugarbeet sucrose is found in

a) Stem b) Leaf

c) Flower d) Root

1595. In sugarbeet sucrose is found in

a) Pod corn b) Dent corn

c) Flint corn d) Pop corn

1596. . Which cereal is non-tillering in nature?

a) Rice b) Wheat

c) Maize d) Barley

1597. Percentage of edible oil in rice bran

a) 5-8% b) 35-40%

c) 25-28% d) 18-20%

1598. Available K less than 100 kg ha^{-1} in soil is ________

a) Low b) High

c) Medium d) Neutral

1599. Nitrification is a ________ process.

a) Physical b) Chemical

c) Biological d) Physic-chemical

1600. In humid regions, availability of precipitation / annum (mm) is

a) 750-1000 b) 1000-1500

c) 500-750 d) >1500

1601. Total water resources of the earth is about

a) 440 m ha m b) 1086 million km^3

c) 1,384 million km^3 d) None

1602. In semi-arid regions, availability of rainfall / annum (mm) is

a) 200-400
b) 250-500
c) 300-500
d) >500

1603. On an average, ________ days is wet days in a year in our country

a) 130
b) 145
c) 155
d) 170

1604. Generally depth of water is expressed in

a) Sq.cm
b) Sq.inch
c) cm
d) m

1605. In India, out of 143 m ha of total cultivated area ________ m ha are rainfed.

a) 101
b) 98.6
c) 87.6
d) None

1606. HYV's were introduced during the mid of ________ in India.

a) 60's
b) 70's
c) 80's
d) 50's

1607. Multi-storied cropping system is commonly practiced in

a) Karnataka & Kerala
b) Bihar & U P
c) Punjab & Rajasthan
d) Tamil Nadu & A P

1608. The most effective cropping system for returning mineral elements to the soil is

a) Mono cropping
b) Double cropping
c) Relay cropping
d) Crop rotation

1609. Which of the following is the prerequisite for multiple cropping ?

a) Availability of HYV seeds
b) Irrigation facility
c) Intensive use of fertilizers
d) None

1610 Specialized farming is one in which

a) A single crop contributes less than 50% to total production
b) A single crop contributes more than 50% to total production
c) More than three crops are taken on same piece of land per unit time
d) None

1611. Temporal complementarity in intercropping results from
   a) The growth patterns of the component crops differing in time
   b) The growth patterns of the component crops differing in height
   c) Differences in yield
   d) Differences in cost of cultivation

1612. The multiple and relay cropping programmes are similar in all respects except ________
   a) The pattern of raising preceding crops differ
   b) The pattern of raising successive crops differ
   c) Both
   d) None

1613. The term 'Cropping scheme' refers to
   a) Allotment of crops to the pieces of land on a farm
   b) Allotment of crop rotation to the pieces of land in a farm
   c) Planning of schedules for different farm operations
   d) Planning and budgeting of crop production in a farm

1614. A non-host crop used to make nematodes to waste their infectional potential is ________
   a) Decoy crop   b) Trap crop
   c) Relay crop   d) Strip crop

1615. Permaculture is an
   a) SRI   b) Integrated farming
   c) Multiple cropping   d) None

1616. Reduced tillage is under
   a) Zero tillage   b) Permaculture
   c) Conservation Agriculture   d) None

1617. SRI is a farmers' developed
   a) Green revolution with high inputs
   b) Methodology
   c) Both
   d) None

1618. Among the millets, ________ is highly drought resistant crop.

a) Kodo millet
b) Finger millet
c) Italian millet
d) Sorghum

1619. The variety of potato which possesses the resistance to frost injury is

a) Kufri jyoti
b) Kufri alankar
c) Kufri sheetaman
d) Kufri chandramukhi

1620. The most important crop grown on conserved moisture is

a) Wheat
b) Mustard
c) Bengal gram
d) Barley

1621. Among the following, which crop is more prone to frost and cold injury?

a) Toria
b) Sarson
c) Banarasi Rai
d) Pahadi Rai

1622. The fruit of linseed is known as ________

a) seed ball
b) Pod
c) Dutum
d) Siliqua

1623. Which of the following causes pungency of mustard oil?

a) Phenols
b) Amino acids
c) Glucosinolates
d) Erucic acid

1624. Which city is known as Agriculture capital of India?

a) New Delhi (Pusa)
b) Coimbatore
c) Hyderabad (Rajendranagar)
d) Mysore

1625. Which state has more number of agricultural universities?

a) UP
b) MP
c) AP
d) Maharastra

1626. Jute Agricultural Research Institute is located at

a) Kolkata
b) Barrackpore
c) Namkum
d) Lucknow

1627. When mungbeans are allowed to sprout, which vitamin is synthesized?

a) Vit A
b) Vit B
c) Vit C
d) Vit E

1628. Dr. V. Kurian is associated with

a) Green revolution
b) Yellow revolution
c) Brown revolution
d) White revolution

1629. Agriculture Price Commission was set up in the year of

a) 1955
b) 1965
c) 1975
d) 1985

1630. Optimum seed rate of lentil varies from ________ kg ha^{-1}.

a) 15-25
b) 30-40
c) 40-50
d) 50-60

1631. The scientific name of field pea is

a) *Pisum sativum* var. *hortense*
b) *Pisum sativum* var. *leguminosarum*
c) *Pisum sativum* var. *esculentum* d)
Pisum sativum var. *arvense*

1632. *Carthamus tinctorius* is the botanical name of

a) Niger
b) Sunflower
c) Safflower
d) Castor

1633. The ratio of cotton seed to lint is

a) 1 : 2
b) 1 : 3
c) 2 : 1
d) 3 : 1

1634. Sesame belongs to the family

a) *Malvaceae*
b) *Pedaliaceae*
c) *Cucurbitaceae*
d) *Asteraceae*

1635. Topping in tobacco means a process of

a) Removal of leaves
b) Burning of leaves
c) Removal of all buds present in the axil of leaves
d) Removal of terminal bud

1636. Early maturing variety of potato is

a) Kufri Pukhraj
b) Kufri Badshah
c) Kufri Lalima
d) HPS-1/13

1637. B-carotene/VIT-A rich variety of sweet potato
   a) Kamala Sundari   b) Sree Nandini
   c) Sree Vardhini   d) Sree Bhadra

1638. Bidhan kusum is a variety of
   a) Safflower   b) Elephant foot yam
   c) Rice   d) Potato

1639. The most serious problem of sweet potato is
   a) Sweet potato feathery mottle virus
   b) sweet potato weevil
   c) Sweet potato scab
   d) Black rot

1640. Sweet potato cuttings are ideal for higher production
   a) Apical vine   b) Middle vine
   c) Bottom vine   d) All of these

1641. Non Polar herbicides are
   a) More soluble in water   b) Less soluble in water
   c) Not soluble in water   d) None of these

1642. Glyphosate 71 SG follows
   a) Symplastic translocation   a) Apoplastic translocation
   b) Both a & b   c) None of these

1643. Paraquat dichloride 24 SL follows
   a) Symplastic translocation   b) Apoplastic translocation
   c) Both a & b   d) None of these

1644. Herbicide selectivity is a phenomenon of
   a) Killing target plants only without or less affecting crop plants
   b) Safe to crop plants
   c) Both
   d) None

1645. The site of action of Triazine group
   a) Photosystem II Site I   b) Photosystem II Site II
   c) Photosystem II Site III   d) None of these

1646. The site of action of Diphenyl ether group

a) Photosynthesis inhibition b) Pigment synthesis

c) Fatty acid Synthesis d) None of these

1647. The site of action of Sulfonyl urea group

a) Photosynthesis inhibition b) Pigment synthesis

c) Acetolactate Synthesis d) None of these

1648. *Phule Pragati* is a variety of

a) Toria b) Rapeseed

c) Sunflower d) Groundnut

1649. Chocolate is responsible for nerve responsive and making energy due to

a) Diketone Pongamol b) Theobromin and Kyaphin

c) Azulene d) Celapanin

1650. *Sweta* is a variety of

a) Toria b) Rapeseed

c) Linseed d) Black mustard

1651. PAC-361 is a variety of

a) Safflower b) Sunflower

c) Linseed d) Black mustard

1652. High % of this component is characteristic of Sunflower oil

a) Linolenic acid b) Oleic acid

c) Palmitic acid d) Stearic acid

1653. Dyes are made from the following plant part of safflower

a) Tender shoots b) Seeds

c) Flowers d) Roots

1654. Hybrid rice seed (F_1) is produced by crossing between ___________ lines.

a) CMS and Maintainer b) CMS and Restorer

c) Maintainer and Restorer d) CMS and Hybrid

1655. The first super-fince grained Basmati type aromatic rice hybrid developed in India is ________

a) APHR 1 b) KRH 2

c) Pusa RH 10 d) CNHR 3

1656. The shape of shelter belt is

a) Square b) Rectangular
c) Round d) Triangle

1657. Design of on-farm research is based on

a) Farmer's designed
b) Researcher designe
c) Researcher implemented
d) Farmer's designed & researcher implemented

1658. Integrated farming system is

a) Holistic approach
b) Crop specific approach
c) Livestock based approach
d) Market demand enterprise approach

1659. Identified technologies could be adopted through

a) Verification trials b) Research trial
c) On farm trial d) On station trial

1660 White Yorkshire is a breed of

a) Rabbit b) Pig
c) Cow d) Rat

1661. Black Bengal is a breed of

a) Goat b) Sheep
c) Cow d) Buffalo

1662. Australop is a breed of

a) Goat b) Duck
c) Poultry d) Dog

1663. Zafarabadi is a breed of

a) Buffaloes b) Cow
c) Rabbit d) Sheep

1664. *Philosamia ricini* is the scientific name of

a) Quail *b) Eri* Silk Worm
c) Duck d) Hen

1665. *Apis cerana indica* is better acclamatised to Higher altitude of

a) Himalayan region
b) Plain region
c) Coastal region
d) Peninsular region

1666. Float dome type is a type of

a) Phospho-compost plant
b) Bio-gas plant
c) Vermicompost plant
d) None of these

1667. The best suited animal component for upland farming system is

a) Goat
b) Buffalo
c) Cow
d) Turkey

1668. Vermiculture technology was initiated by

a) Kerala Agricultural University
b) Assam Agricultural University
c) University of Agricultural Sciences, Bangalore
d) TNAU, Coimbator

1669. *Trichogramma japonicum* is effective against

a) Aphid
b) Whorl maggot
c) Ear head bug
d) Stem borer

1670. Polyculture fingerlings required for 0.04ha of ponded water is ________ numbers.

a) 400
b) 4000
c) 6000
d) 40000

1671. The most suitable horticultural crop in arid land farmlng system is

a) Mango
b) Guava
c) Litchi
d) Ber

1672. For optimum production of bio-gas C:N ratio should be maintained as

a) 10 : 1
b) 20 : 1
c) 30 : 1
d) 40 : 1

1673. The objective of subsistence farming is to produce

a) Fibre and food crops for family and market
b) Essential food crops for family only
c) Essential food crops for family and market
d) Vegetable crops for family only

1674. Inclusion of *Stylosanthes hamata* as a pasture legume in rotation with castor or sorghum is referred as

a) Intensive cropping
b) Sequence cropping
c) Relay cropping
d) Ley farming

1675. Rice Bran Oil contains

a) Cyanolipids
b) Cyclopentenyl fatty acid
c) Gramma Oryzanol
d) Friol oil

1676. Root depth of potato considered for irrigation is

a) 15 cm
b) 30 cm
c) 45 cm
d) 60 cm

1677. Green fodder of oat contains ________ % protein.

a) 5-7
b) 15-20
c) 10-12
d) 20-23

1678. Scientific name of Witch weed is

a) *Leucas aspera*
b) *Striga asiatica*
c) *Vicia sativa*
d) *Parthenium hysterophorus*

1679. *Cuscuta reflexa* is commonly known as ________

a) Jimson weed
b) Mutha gash
c) Dodder
d) Horse weed

1680. Safener is used with herbicides to

a) Protect the crop
b) Increase activity
c) Increase phytotoxicity
d) None

1681. Acetamide herbicide Butachlor is commonly used in _________

a) Cotton
b) Tea
c) Paddy
d) Jute

1682. The ideal tillage method for semi-arid tropic region is ________

a) Repeated deep tillage
b) Blind tillage
c) Zero tillage
d) Reduced tillage

1683. Application of nitrogen in pulses at the time of sowing is known as:

a) Basic dose
b) Starter dose
c) Additional dose
d) Synergistic dose

1684. The highest nitrogen content among the commonly used nitrogenous fertilizers is embodied by

a) Ammonium nitrate
b) Ammonium sulphate
c) Calcium ammonium nitrate
d) Urea

1685. Symbiotic association of fungal hyphae with plant roots is known as _____

a) Lichen
b) Co-parasitism
c) Mutual infection
d) Mycorrhiza

1686. Inherent capacity of soil to supply nutrients in adequate amount and suitable proportions is called as ________

a) Fertility index
b) Productivity index
c) Soil fertility
d) Soil productivity

1687. Among the following, acid forming fertilizer is ________

a) Urea
b) Sodium nitrate
c) Potassium chloride
d) Calcium ammonium nitrate

1688. Maximum amount of P_2O_5 is in ________

a) DAP
b) DCP
c) SSP
d) TSP

1689. The term 'lancing' is related with ________ crop

a) Sugarcane
b) Tobacco
c) Isabgol
d) Opium poppy

1690. The best green manuring crop for rice field is ________

a) Guar
b) Dhaincha
c) Sunnhemp
d) Berseem

1691. Fertilizer which supplies three essential plant nutrients is

a) DAP
b) SSP
c) MOP
d) SOP

1692. ________ fertilizer is an important component of precision farming

a) Straight
b) Specialty
c) Complex
d) Mixed

1693. Horticulture based farming system is generally found in

a) Himachal Pradesh, Jammu & Kashmir and Sikkim

b) W.B., Bihar, Odisha

c) Madhya Pradesh

d) Uttar Pradesh

1694. "Permaculture" word was coined by

a) Fukuoka
b) Rudolf Steiner
c) Albert Howard
d) Bill Mollison

1695. Animal based farming is most common in

a) West Bengal
b) Odisha
c) Rajasthan
d) Delhi

1696. In stale seedbed technique weeds are killed by.

a) Flaming
b) Light cultivation,
c) Tillage
d) All of these

1697. The desired varieties of economically useful crops are raised by

a) Vernalisation
b) Mutation
c) Natural selection
d) Hybridisation

1698. High-yielding varieties of wheat were primarily developed by Indian scientist by crossing- breeding traditional varieties with

a) American varieties
b) Mexican varieties
c) European varieties
d) African varieties

1699. A plant breeder wants to develop a disease resistant variety. What should he do first?

a) Hybridisation
b) Mutation
c) Selection
d) Production of crop

1700. Selection of homozygous plant is

a) Pure line selection
b) Mass selection
c) Mixed selection
d) Introduction

1701. If wheat field is inoculated with Rhizobium

a) Soil will become nitrogen rich
b) No effect on soil nitrogen
c) Soil will be depleted of nitrogen
d) Soil will become rich in calcium

1702. In mushroom cultivation, spawn is

a) Compost b) Button stage

c) Vegetative mycelium d) Harvested mushroom

1703. Insect deterrent chemical extracted from neem (margosa) is

a) Thiocarbamate b) Pyrethroid

c) Azadirachtin d) Allethrin

1704. Leaves of plant used as biofertilizer belong to

a) Hibiscus b) Mango

c) Anabeana d) Azolla

1705. Leghaemoglobin is found in

a) Root nodules of legumes b) Mycorrhiza

c) Coralloid roots d) Cyanobacteria

1706. Leghaemoglobin takes part in

a) Energy release b) Stimulating growth of *Rhizobium*

c) N_2 absorption d) Protecting nitrogenase

1707. Leguminous plants are able to fix atmospheric nitrogen through symbiotic process. Which is not true about it?

a) Leghaemoglobin scavenges oxygen and is pinkish in colour

b) Nitrogenase is insensitive to Oxygen

c) Nodules act as site for nitrogen fixation

d) Enzyme nitrogenase catalyses conversion at atmospheric nitrogen to NH_3

1708. Leghaemoglobin occurs in

a) Coralloid root b) BGA

c) Around bacteriods d) Mycorrhiza

1709. Main sources of biofertilizers are

a) Bacteria b) Cyanobacteria

c) Fungi d) All of the above

1710 A biofertilizer which involves a pteridophyte host is

a) Rhizobium b) Anabaena

c) Clostridium d) Azotobacter

1711. A fern commonly inoculated to paddy fields is

a) Azolla b) Marsilea

c) Salvinia d) Anabaena

1712. A free living nitrogen fixing cyanobacterium which can also form symbiotic association with Azolla is

a) Nostoc b) Anabaena

c) Tolypothrix d) Gleocapsa

1713. A medicine for bronchitis is got from

a) Rauwolfia serperntina b) Curcuma longa

c) Adhatoda vasica d) Hemidesmus indicus

1714. A nitrogen fixing blue green alga is

a) Ulothrix b) Spirogyra

c) Anabaena d) Rhizobium

1715. A piosnous mushroom among the following is

a) Agaricus bisporus b) Morchella esculenta

c) Hydnum sp. d) Amanita sp.

1716. A plant effective in ensuring safe delivery ans prevent abortions is

a) Azadirachta b) Ocimum

c) Adhatoda d) Asparagus

1717. Aquatic fern which is an excellent biofertilizer

a) Salvinia b) Azolla

c) Marsilea d) Pteridium

1718. Azolla is used as biofertilizer because it

a) Multiplies very fast to produce massive biomass

b) Has association of nitrogen fixing *Rhizobium*

c) Has association of nitrogen fixing cyanobacteria

d) Has association of mycorrhiza

1719. *Azotobacter* and *Bacillus polymyxa* are

a) Decomposers b) Nonsymbiotic N fixers

c) Symbiotic N fixers d) pathogenic bacteria

1720. Biofertilizers include

a) Cowdung manure and farmyard waste

b) A quick growing crop ploughed back

c) BGA / Anabeana and Azolla

d) All the above

1721. Major glycoside present in leaf juice of Aloe vera is

a) Emodin b) Aloe emodin

c) Barbaloin d) Aloin

1722. Most famous nitrogen fixing bacterium / biofertililzer is

a) *Nitrobacter* b) *Nitrosomonas*

c) *Nitrococcus* d) *Rhizobium*

1723. Most suitable fertilizer of paddy fields is

a) Mycorrhiza

b) *Azotobacter* and *Clostridium*

c) Symbiotic and nonsymbiotic cyanobacteria

d) *Rhizobium*

1724. Mushroom cultivation consists of steps as

a) Spawning - Composting - Harvesting - casing

b) Casing - Composting - Harvesting - Spawning

c) Cropping - Casing - Spawning - Composting

d) Composting - Spawning - Casing - Harvesting

1725. Mycorrhiza is helpful in

a) Synthesis of food

b) Getting nutrients from soil

c) Providing resistance against different regulators

d) Increase the fertility of soil

1726. Mycorrhiza is symbiotic association between

a) Fungus and gymnosperm stem

b) Fungus with angiosperm leaves

c) Fungus with legume fruits

d) Fungus with Gymnosperm and angiosperm roots

1727. Nitrogen fixation is

a) Nitrogen - Ammonia
b) Nitrogen - Nitrates
c) Nitrogen - Amino acids
d) Both a and b

1728. *Ocimum sanctum* belongs to the family

a) *Meliaceae*
b) *Acanthaceae*
c) *Liliaceae*
d) *Lamiaceae*

1729. Red pigment (leghaemoglobin) having affinity for oxygen is present in the roots of

a) Mustard
b) Soyabean
c) Carrot
d) Radish

1730. *Rhizobium* induced root nodules are internally pinkish due to

a) Carotene
b) Leghaemoglobin
c) Haemoglobin
d) Xanthophyll

1731. Which is correct?

a) Legumes fix nitrogen through bacteria in their leaves
b) Legumes fix nitrogen through bacteria in their roots
c) Legumes fix nitrogen independent of bacteria
d) Legumes do not fix nitrogen

1732. Which of the following can use molecular nitrogen as nutrient?

a) *Methanomonas*
b) *Mucor*
c) *Rhizobium*
d) *Spirogyra*

1733. Which of the following is free living aerobic and nonphotosynthetic nitrogen fixing bacterium?

a) *Rhizobium*
b) *Nostoc*
c) *Azospirillum*
d) *Azotobacter*

1734. Which one is a biofertilizer?

a) NPK mixture
b) Rhizobia in legume roots
c) Rhizobia in farmyard manure
d) Green manure

1735. Which one is an edible fungus

a) *Mucor*
b) *Agaricus*
c) *Penicillium*
d) *Rhizopus*

1736. Which one is medicinally exploited in Adhatoda zeylanica ?

a) Leaves
b) Roots
c) Flowers
d) All the above

1737. Which one is used in cosmetics

a) Calotropis
b) Costus speciosum
c) Chlorophytum borivilium
d) Aloe vera

1738. Which one of the following is free living nitrogen fixing bacterium

a) Azotobacter
b) Anabaena azollae
c) Pseudomonas
d) Cyanobacterium

1739. The science concerned with vegetable culture is called

a) Floriculture
b) Olericulture
c) Horticulture
d) Agriculture

1740. Although a deficiency of any one of the elements listed may result in chlorosis, only one of these elements is an element found in chlorophyll. Which is it?

a) Zinc
b) Iron
c) Magnesium
d) Chloride

1741. Which of the following elements is not present in a nitrogenous base?

a) Hydrogen
b) Carbon
c) Phosphorus
d) Nitrogen

1742. A water-fern, which is used as a green manure in rice fields, is

a) *Salvinia*
b) *Mucor*
c) Aspergillus
d) *Azolla*

1743. Which one among the following chemicals is used for causing defoliation of forest trees?

a) Posphon D
b) Malic hydrazide
c) 2, 4-D
d) Amo 1618

1744. Bioherbicides have been recommended

a) To prevent eco-degradation
b) Because of their ready availability
c) Because of their cheap rates
d) Because of their abundance

1745. The most important weed against which eradication measures would be taken on warfootings is

a) *Eichhornia*
b) *Dactylopius*
c) *Parthenium*
d) *Ageratum*

1746. Water logging of soil makes it physiologically dry because

a) This condition does not allow the capillary force to work

b) This condition does not allow oxygen to enter the soil

c) Both (a) and (b)

d) None of these

1747. The process by which nutrient chemicals or contaminants are dissolved and carried away by water, or are moved into a lower layer of soil

a) Mulching
b) desertification
c) Incineration
d) Leaching

1748. First bioinsecticide developed commercial scale was

a) Quinine
b) DDT
c) Organophosphate
d) Sporeine

1749. Which of the following elements is almost not essential for plants?

a) Ca
b) Mo
c) Zn
d) Na

1750. Major food crops of the world belong to family

a) Leguminosae
b) Gramineae
c) Solanaceae
d) Cruciferae

1751. VAM is a ________

a) Virus
b) Alga
c) Fungus
d) Bacterium

1752. *Azolla* is used as biofertilizer as it has ________

a) Azotobacter
b) Cyanobacteria
c) Mycorrhiza
d) Large quantity of humus

1753. The crop which has the highest cultivated area in the world is ________

a) Rice
b) Wheat
c) *Bajra*
d) Barley

1754. There are ________ cultivated species of *Oryza*

a) 8
b) 6
c) 4
d) 2

1755. Percentage of edible oil in soybean is ________

a) 10-12%	b) 13-15%

c) 16-18%	d) 20-22%

1756. The idea of randomization was put forwarded by ________

a) Chapman	b) Gomez

c) Sukathmae	d) Fisher

1757. Poor man's timber is ________

a) Mango wood	b) Jackfruit wood

c) Babla wood	d) Bamboo

1758. Collego is a ________

a) Mycoherbicide	b) Chemical herbicide

c) Both	d) None

1759. Which of rice sub-species is having photo-insensitive nature?

a) Indica	b) Japonica

c) Javanica	d) Both b & c

1760 Which part of panicle later turned into hull?

a) lemma	b) palea

c) both	d) none

1761. First dwarf variety of the world which also considered as **miracle rice** is

a) IR 8	b) Jaya

c) ADT 27	d) TN 1

1762. Country which is pioneer in hybrid rice is

a) China	b) India

c) Philippines	d) Japan

1763. Sowing and harvesting months of *Aus* rice are

a) May-June and September-October

b) June-July and November-December

c) November-December and March-April

d) None of these

1764. Major loss of applied N in low condition is due to

a) Denitrification	b) Leaching

c) Ammonia volatilization	d) None of these

1765. Seeds required to sow one hectare of land under broadcast puddle condition of rice is

a) 60 kg/ha b) 80 kg/ha

c) 100 kg/ha d) 125 kg/ha

1766. Which is the best fertilizer for basal application of paddy puddle field

a) Ammonium sulphate b) Urea

c) DAP d) All

1767. In India which type of maize is widely grown

a) Dent corn b) Flint corn

c) Sweet corn d) Waxy corn

1768. White bud of maize symptom is due to deficiency of

a) N b) P

c) K d) Zn

1769. Optimum depth of sowing of Wheat seeds is

a) 3-5 cm b) 1-2 cm

c) 8-10 cm d) 5-10 cm

1770. *Chorchorus olitorius* (*Mitha Paat*) is origainated from

a) India b) Africa

c) Indo-Burma d) USA

1771. One hectare metre is equal to ________ m^3 of water

a) 100 b) 1000

c) 10000 d) 100000

1772. Pulse crop grown for grain, fodder and green manure is

a) Cowpea b) Redgram

c) Gram d) Lentil

1773. Factor influencing the flow of water under saturated condition

a) Hydraulic conductivity

b) Hydraulic force driving the water through soil

c) Both a and b

d) None of these

1774. Water in excess of field capacity is termed as ____________ water
a) Gravitational b) Capillary
c) Available d) Both b & c

1775. Which crop stage is most susceptible for water logging
a) Seedling b) Vegetative
c) Flowering d) Maturity

1776. Actual ET depends on
a) Climate b) Soil moisture
c) Both d) None

1777. Factor influencing effective rainfall is
a) Amount/intensity of rainfall
b) Soil tilth
c) Moisture storage capacity of soil
d) All of these

1778. Major source of irrigation in India is
a) Canal b) Tank
c) Well d) Others

1779. Water use efficiency of rice is around ________ kg/ha mm
a) 1.0 b) 2.0
c) 3.0 d) 4.0

1780. Which phase of sugarcane is critical for water
a) Formative b) Grand growth
c) Maturity d) None

1781. Temperature requirement for grain filling of wheat is ________ °C
a) 20-25 b) 20-23
c) 23-25 d) 16-20

1782. Contribution of flag leaf in photosynthates is about
a) 52% b) 40%
c) 35% d) 12%

1783. Which one of the following crops is most sensitive to both excess moisture and drought
a) Direct seeded rice b) Maize
c) Sunflower d) Sorghum

1784. Atrazine used as an antitranspirant

a) Reduces the growth of the crop

b) Does not reflect from plant leaf surface

c) Affects the closure and opening of stomata

d) Forms thin layer on the leaf surface

1785. Productivity of an intercrop per unit area of ground compared with that expected from sole crop sown in the same proportions is termed as

a) Land equivalent ratio b) Competitive ratio

c) Land equivalent coefficient d) Crop performance ratio

1786. The process of replacement of one atom by another atom of similar size, in a crystal lattice of a soil clay without disrupting or changing the crystal structure of a mineral is termed as

a) Ion exchange b) Isomerism

c) Isomorphic substitution d) Polymorphism

1787. The element which does not enter into permanent organic combination in plants is

a) N b) P

c) Mg d) K

1788. The major organic cementing agent in soil aggregate formation is

a) Lipid b) Proteins

c) Polysacchaides d) Organic acids

1789. The cohensiveness and tenderness of cooked rice depends largely on

a) The percentage of amylase

b) The percentage of amylopectin

c) The proportion of amylase and protein

d) The proportion of amylase and amylopectin

1790. Which of the clay minerals are found most in agriculture soils

a) Cyclosilicates b) Phyllosilicates

c) Sorosilicates d) Nesosilicates

1791. Which one of the following solid particles will form a soil suspension exhibiting high colloidal properties

a) Kaolinite b) Vermiculite

c) Illite d) Hematite

1792. Which one of the following solid particles will form a soil suspension exhibiting high colloidal properties

a) Zn b) Fe

c) Cu d) Mn

1793. Sorghum crop is considered as camel crop because of

a) Deep root system b) Resistant to drought

c) Shallow root system d) Nutrient exhaustiveness

1794. V-shaped furrow is opened by

a) Disc plough b) MB plough

c) Country plough d) Chisel plough

1795. Long term persistent herbicide is

a) Pendimithalin b) 2,4-D

c) Paraquat d) Glyphosate

1796. Composted manure is produced from

a) Farmyard manure and green manure

b) Farm refuse and household refuse

c) Organic remains of biogas plants

d) Rotten vegetables and animal refuse

1797. Norin-l0 gene from Japan is a

a) Dwarf gene of wheat b) Dwarf gene of rice

c) Dwarf gene of maize d) Disease resistant gene of rice

1798. Aims of plant breeding are to produce

a) Disease-free varieties b) High-yielding varieties

c) Early-maturing varieties d) All of the above

1799. Growing of two or more crops simultaneously on the same piece of land is called

a) Mixed cropping b) Mixed farming

c) Intercropping d) Fanning

1800. The Mexican dwarf wheat variety was developed by

a) Swaminathan b) Borlaug

c) Watson d) Khush

1801. Imperial Council of Agricultural Research was established in

a) 1926 b) 1928

c) 1929 d) 1933

1802. CTCRI is situated in

a) Uttarakhand b) Andhra Pradesh

c) Kerala d) Uttar Pradesh

1803. Indian Institute for Sugercane Research is situated in

a) Tamil Nadu b) Uttar Pradesh

c) Madhya Pradesh d) Karnataka

1804. Research institute for Lac is situated at

a) Ranchi b) Lucknow

c) Varanasi d) Kanpur

1805. Potato Breeding and Certification Station is situated in

a) Himachal Pradesh b) Uttarakhand

c) Kashmir d) Uttar Pradesh

1806. Research Institute in Kerala stand for

a) Fodder b) spices

c) Sugarcane d) Horticulture

1807. CPCRI is situated in

a) Kerala b) Karnataka

c) Tamil Nadu d) Telengana

1808. CAZRI is situated in

a) Gujrat b) Rajasthan

c) Karnataka d) UP

1809. Institute for cotton technology is situated in

a) Nagpur b) Mumbai

c) Bhopal d) Ludhiana

1810. Indian Institute of Soil Science is situated in

a) Tamil Nadu b) Uttar Pradesh

c) Uttarakhand d) Madhya Pradesh

1811. Undesirable, troublesome weed difficult to control

a) Noxious b) Objectionable

c) Both d) None

1812. Cuscuta is

a) Total root parasite b) Total stem parasite

c) Semi root parasite d) Semi stem parasite

1813. Alachlor belongs to the group

a) Amide b) Dinitroaniline

c) Triazine d) Organophosphorus

1814. What is the trade name of Fenoxaprop ethyl?

a) Turga super b) Whip super

c) Sunrise d) Ronstar

1815. Propanil is a ________ herbicide

a) Contact, selective b) Systemic,non-selective

c) Contact, non-selective d) Systemic, selective

1816. The average annual crops loss due weed is

a) 15-20% b) 20-30%

c) 30-40% d) 40-50%

1817. Transpiration ratio of *Cynodon dactylon* (Bermuda grass) is

a) 613 b) 713

c) 813 d) 913

1818. Trade name of Clodinafop Propargyl

a) Select b) Topik

c) Crop star d) Topstar

1819. 2,4-D was first time tested in India in the year

a) 1942 b) 1944

c) 1946 d) 1948

1820. *Octotoma scrabipennis* is used to control

a) Parthenium b) *Alternanthera*

c) *Cynodon dactylon* d) *Lantana camara*

1821. Cubic metre is equal to

a) 100 lit b) 1000 lit

c) 10000 lit d) 100000 lit

1822. For sandy soil

a) FC>ME b) FC<ME

c) FC=ME d) None

1823. Higher the clay content of a soil, ______ its hyhygroscopic coefficient

a) Higher b) Lower

c) Near to equal d) None

1824. Water conveyance efficiency denotes

a) Water delivered to the field/Water needed in the root zone

b) Water needed in the root zone/ Water delivered from the source

c) Water delivered to the field/ Water delivered from the source

d) Water needed in the root zone/ Water delivered to the field

1825. Wilting coefficient implies

a) Hyhygroscopic coefficient/1.8 b) Moisture equivalent/0.68

c) Hyhygroscopic coefficient/0.68 d) Moisture equivalent/0.8

1826. For drought resistant crops, allowed depletion of available soil moisture is

a) 40% b) 50%

c) 60% d) 70%

1827. IW/CPE ratio for barseem

a) 0.4 b) 0.6

c) 1.0 d) 1.2

1828. Depth of ploughing for deep rooted crops is .... cm

a) 15-20 b) 20-25

c) 25-30 d) 30-35

1829. Furrow irrigation is not suitable for which soil?

a) Sandy b) Clay

c) Loamy d) All

1830. BBF (Broad-bed and furrow) system is applicable for which soil?

a) Sandy b) Clay

c) Loamy d) All

1831. Bed width for BBF system is ________ cm.

a) 110-130 b) 120-150

c) 130-160 d) 140-170

1832. Pressure of sprinkler irrigation is ________ kg/cm^2.

a) 0.5-2.5 b) 1.5-2.5

c) 2.5-4.5 d) 3.5-4.5

1833. Cablegation is atype of ________ irrigation.

a) Surface b) Sub-surface

c) Both d) None

1834. Conveyance loss in surge irrigation is ________ %

a) <15 b) >15

c) <20 d) >20

1835. Shallow and small furrows are known as ________

a) Dike b) Corrugation

c) Surge d) Bed

1836. Water use efficiency of rice is ________

a) 2.5 b) 3.1

c) 3.7 d) 4.7

1837. Concept of PET was given by ________

a) Sharma and Dastane b) Thornthwaite

c) Penmann d) Parihar

1838. Which one is the best method for estimating soil moisture?

a) Tensiometer b) Gypsum block

c) Neutron probe d) Pressure plate

1839. Available water concept was given in the year

a) 1978 b) 1981

c) 1984 d) 1987

1840. Sunken screen evaporimeter was given by

a) Sharma and Dastane b) Penmann

c) Thornthwaite d) Both b and c

1841. Porosity implies

a) [1-(BD/PD)]*100
b) [1-(PD/BD)]*100
c) [(BD/PD)-1]*100
d) [(PD/BD)-1]*100

1842. Void ratio = ?

a) 1-(BD/PD)
b) 1-(PD/BD)
c) (BD/PD)-1
d) (PD/BD)-1

1843. Bulk density =?

a) Mass wetness/volume wetness
b) Volume wetness/mass wetness
c) Mass wetness*volume wetness
d) Volume wetness*mass wetness

1844. Porosity/(1-porosity) = ?

a) Bulk density
b) Air filled porosity
c) Void ratio
d) Particle density

1845. Which one is called "water of cohesion"?

a) Gravitational water
b) Capillary water
c) Hygroscopic water
d) All

1846. At PWP, wilting symptom lasts for ________ hours.

a) 12
b) 24
c) 48
d) 72

1847. Formula of relative leaf water content (RLWC) is ________

a) (Fresh weight-dry weight)/turgid weight
b) (Fresh weight-dry weight)/(turgid weight-fresh weight)
c) (Turgid weight-fresh weight)/ (Fresh weight-dry weight)
d) (Fresh weight-dry weight)/(turgid weight-dry weight)

1848. Severe stress occurs when relative leaf water content drops by _____%

a) 5
b) 10
c) 15
d) 20

1849. Which law relates flow rate of liquid with the driving force?

a) Poiseuille's law
b) Darcy's law
c) Both
d) None

1850. Normally diffusion pressure deficit of a cell is

a) OP+TP-WP
b) OP+WP-TP
c) OP-TP
d) TP-OP

1851. Which one is called King of cereals?

a) Rice b) Wheat

c) Maize d) Oat

1852. King of temperate fruits

a) Apple b) Pear

c) Peach d) Apricot

1853. Queen of vegetables

a) Onion b) Cabbage

c) Potato d) Pointed gourd

1854. Poor man's meat

a) Cowpea b) Soybean

c) Mushroom d) Potato

1855. Famine reserves

a) Potato b) Rice

c) Millets d) Pulses

1856. Vegetarian meat

a) Soybean b) Cowpea

c) Potato d) Mushroom

1857. Food of God

a) Apple b) Cocoa

c) Cashew d) Saffron

1858. Queen of oilseed

a) Groundnut b) Sesame

c) Soybean d) Linseed

1859. King of oilseed

a) Mustard b) Groundnut

c) Soybean d) Sunflower

1860. Queen of fodder crop

a) Barseem b) Lucern

c) Cowpea d) Napier grass

1861. King of fodder crop

a) Barseem
b) Cowpea
c) Napier grass
d) Oat

1862. Poor man's substitute for ghee

a) Mustard
b) Soybean
c) Sesame
d) Groundnut

1863. Queen of cereals

a) Rice
b) Maize
c) Sorghum
d) Oat

1864. Vegetable meat

a) Soybean
b) Mushroom
c) Potato
d) Cowpea

1865. Camel crop

a) Date palm
b) Pearl millet
c) Sorghum
d) Finger millet

1866 Poor man's friend

a) Sesame
b) Potato
c) Soybean
d) Cowpea

1867 Poor's men fruits

a) Banana
b) Ber
c) Mango
d) Guava

1868. King of arid fruit

a) Date palm
b) Guava
c) Ber
d) Banana

1869. Green gold

a) Opium
b) Bamboo
c) Soybean
d) Tea

1870. White gold

a) Wheat
b) Cotton
c) Rice
d) Silk

1871. Century plant

a) Date palm
b) Coconut
c) Arecanut
d) Ber

1872. Queen of beverage

a) Tea
b) Coffee
c) Lemon juice
d) Coconut

1873. Queen of flower

a) Rose
b) Gladiolus
c) Dahlia
d) Night queen

1874. Queen of spices

a) Cinnamon
b) Cardamom
c) Clove
d) Fenugreek

1875. Queen of pulses

a) Lentil
b) Chickpea
c) Pea
d) Green gram

1876. King of pulses

a) Lentil
b) Chickpea
c) Red gram
d) Black gram

1877. Queen of fruits

a) Banana
b) Litchi
c) Guava
d) Pineapple

1878. King of coarse cereals

a) Sorghum
b) Pearl millet
c) Finger millet
d) Proso millet

1879. Drosophila of crop plants

a) Rice
b) Wheat
c) Maize
d) Soghum

1880. Poor man's food

a) Soybean
b) Sesame
c) Pearl millet
d) Potato

1881. King of weed

a) Dub grass
b) Mutha
c) Congress grass
d) *Alternanthera*

1882. Adam's fig

a) Apple
b) Banana
c) Guava
d) Litchi

1883. King of forest

a) Teak
b) Oak
c) Pine
d) Bamboo

1884. Butter fruit-

a) Almond
b) Apricot
c) Cashew
d) Avocado

1885. Poor man's orange

a) Litchi
b) Carrot
c) Tomato
d) Lime

1886. Poor man's apple

a) Litchi
b) Berry
c) Ber
d) Guava

1887. Love apple

a) Guava
b) Tomato
c) Orange
d) Litchi

1888. Gypsophyla Crop

a) Almond
b) Apricot
c) Cashew
d) Wallnut

1889. King of spices

a) Cinnamon
b) Blackpepper
c) Cardamom
d) Fennel

1890. Wonder crop

a) Dhaincha
b) Soybean
c) Linseed
d) Cowpea

1891. Blanket flower is

a) Dahlia
b) Gaillardia
c) Lotus
d) Rose

1892. Miracle fruit is

a) Pineapple
b) Kiwi
c) Guava
d) Apple

1893. KalpaVriksha is

a) Sal
b) Coconut
c) Date palm
d) Bamboo

1894. Queen of flowers

a) Gladiolus
b) Lotus
c) Rose
d) China rose

1895. Tree of paradise

a) Apple
b) Coconut
c) Banana
d) Date palm

1896. Apple of Paradise

a) Phalsa
b) Banana
c) Apple
d) Date palm

1897. Monkey Jack

a) Jack fruit
b) Phalsa
c) Date palm
d) Apple

1898. Apple of Tropic

a) Phalsa
b) Apple
c) Date palm
d) Guava

1899. King of Arid fruits

a) Phalsa
b) Ber
c) Coconut
d) Date palm

1900. Five Corner fruit

a) Carambola
b) Phalsa
c) Coconut
d) Date palm

1901. Star Apple

a) Jack fruit
b) Litchi
c) Pineapple
d) Phalsa

1902. Chinese Fig

a) Apple
b) Black berry
c) Sapota
d) Ber

1903. Queen of Nut

a) Macadamia nut
b) Pecan nut
c) Brazil nut
d) Walnut

1904. King of Nut

a) Pecan nut
b) Peanut
c) Walnut
d) Cashew nut

1905. Gold mine of Waste land

a) Bamboo
b) Cashew
c) Sesame
d) Pearl millet

1906. Sapodilla plum

a) Coconut
b) Date palm
c) Arecanut
d) Sapota

1907. Fancy Fruit

a) Strawberry
b) Dragon fruit
c) Mandarin
d) Custard apple

1908. Black Plum

a) Date palm
b) Jamun
c) Ber
d) Grape

1909. Indian Black Berry

a) Jamun
b) Grape
c) Date palm
d) Black pepper

1910. Indian Gooseberry

a) Sapota
b) Grape
c) Strawberry
d) Amla

1911. Malacca Tree

a) Sapota b) Apple

c) Anola d) Mango

1912. Wolf Apple

a) Onion b) Custard apple

c) Phalsa d) Tomato

1913. Vilayati Baigan

a) Tomato b) Jack fruit

c) Potato d) Capsicum

1914. Butter Bean

a) Kidney bean b) Lima bean

c) Urd bean d) Broad bean

1915. Egg plant

a) Tomato b) Chilli

c) Brinjal d) Capsicum

1916. Indian Bean-

a) Broad bean b) Snap bean

c) Kidney bean d) Dolichos Bean

1917. Vegetable of 20th century

a) Dolichos Bean b) Winged Bean

c) Potato d) Fig

1918. Vegetable of 21st century

a) Chekurmanis b) Bitter Gourd

c) Snap bean d) Drumstick

1919. Black eyed pea

a) Kidney bean b) Cowpea

c) Dolichos Bean d) Snap bean

1920. Wholesome Food

a) Potato b) Musk Melon

c) Kidney bean d) Apple

1921. King of Annual Flower

a) Rose b) Pensy
c) Daffodil d) Tulip

1922. Potato Bean

a) White yam b) Yam Bean
c) Sweet potato d) Cassava

1923. Misrikand

a) Sweet potato b) Cassava
c) Yam d) Yam Bean

1924. Balsam Pear

a) Bitter Gourd b) Bottle Gourd
c) Ridge Gourd d) Snake Gourd

1925. Bitter Cucumber

a) Bitter gourd b) Pointed gourd
c) Snake Gourd d) Ash gourd

1926. Vegetable of Immense Value

a) Pointed gourd b) Potato
c) Pumpkin d) Cabbage

1927. Swiss Chord

a) Drumstick b) Pointed gourd
c) Palak d) Ladies finger

1928. Horse Radish Tree

a) Tomato b) Drumstick
c) Brinjal d) Potato

1929. Schlor's Tree

a) Ashok b) Pipal
c) Gulmohar d) Sal

1930. Bodhi Tree

a) Banian b) Ashok
c) Segun d) Pipal

1931. Peacock tree

a) Gulmohar b) Maple

c) Pine d) Sourwood

1932. Pot Marigold

a) Rudbeckia b) Dimorphotheca

c) Calendula d) Phlox

1933. Dog Flower

a) Rudbeckia b) *Gurhal*

c) Dimorphotheca d) Antirrhinum

1934. Backbone of America

a) Maize b) Soybean

c) Wheat d) Cotton

1935. Golden fibre

a) Jute b) Mesta

c) Linseed d) Both a and b

1936. Small holder's irrigated crop

a) Date palm b) Coconut

c) Arecanut d) Oil palm

1937. Brown gold

a) Vermicompost b) Dead pupa of silkworm

c) Soil d) Jute

1938. King of vegetables

a) Potato b) Cabbage

c) Okra d) Brinjal

1939. Glory of East

a) Gladiolus b) Rose

c) Crysanthemum d) Tulip

1940. Bio drainage plant

a) Banian tree b) Eucalyptus

c) Coconut d) Mango

1941. Trashing is done in

a) Tobacco b) Sugarcane
c) Cotton d) Banana

1942. Stalking is done in

a) Cotton b) Gram
c) Tea d) Tomato

1943. Rabbing is done in

a) Gram b) Banana
c) Lucerne d) Tobacco

1944. Lopping is done in

a) Lucerne b) Sugarcane
c) Rice d) Sunflower

1945. Standing is done in

a) Tea b) Sugercane
c) Sunflower d) Tomato

1946. Tipping is done in

a) Tobacco b) Tea
c) Gram d) Tomato

1947. Stripping is done in

a) Tobacco b) Sugarcane
c) Jute d) Banana

1948. Tying is done in

a) Sugarcane b) Tea
c) Cotton d) Groundnut

1949. Topping is done in

a) Tobacco b) Gram
c) Tea d) Cotton

1950. De-suckering is done in

a) Banana b) Tobacco
c) Sugarcane d) Sweet potato

1951. Concept of LAI was given in

a) 1938
b) 1948
c) 1958
d) 1968

1952. Concept of LAI was given by

a) Donald
b) Watson
c) Gregory
d) Blackman

1953. LAI implies

a) Leaf area/total existing dry matter
b) Leaf area/dry matter produced in a day
c) Leaf area/ age of crop
d) Leaf area/ground area

1954. Cropping intensity implies

a) (Net cropped area/gross cropped area)*100
b) (Gross cropped area/Net cropped area)*100
c) (Gross cropped area/ No. of crops grown in a year)*100
d) (Net cropped area/ No. of crops grown in a year)*100

1955. Intercropping is beneficial when

a) LER>2
b) LER>1
c) LER>0
d) LER=1

1956. Harvest index implies

a) Grain yield/ straw yield
b) Grain yield/ grain+straw yield
c) Economic yield/biological yield
d) Both b and c

1957. Coefficient of effectiveness implies

a) (Source/sink)*100
b) (Sink/Source)*100
c) (Yield/Source)*100
d) (Yield/Sink)*100

1958. Light use efficiency means

a) Dry matter produced/Cumulative light absorbed
b) Dry matter produced/ light transmitted through the canopy
c) Dry matter produced/ light absorbed in a day
d) Cumulative light absorbed/ Dry matter produced

1959. Absolute growth rate means

a) g of dry matter produced per unit land area
b) g of dry matter produced per unit leaf area
c) g of dry matter produced per day
d) g of dry matter produced per unit land area per day

1960 CGR is expressed in

a) $g/m^2/day$
b) $g/g/m^2$
c) g/g/day
d) $cm/ m^2/day$

1961. Absolute growth rate is expressed in

a) g/g/day
b) g/day
c) g/m^2
d) cm/day

1962. Relative growth rate concept was given by

a) Watson
b) Blackman
c) Donald
d) Gregory

1963. NAR concept was given by

a) Blackman
b) Watson
c) Gregory
d) Donald

1964. NAR is expressed in

a) $g/ m^2/day$
b) $cm/ m^2/day$
c) $g/cm^2/day$
d) g/cm^2

1965. Rice is a ________ limited crop.

a) Sink
b) Source
c) Both
d) None

1966. Source and sink are developed simultaneously in

a) Groundnut
b) Cotton
c) Redgram
d) All

1967. dy/dx= (A-y) C indicates

a) Liebig's law
b) Blackman's law
c) Mitscherlich's law
d) Inverse-yield Nitrogen law

1968. Germination of groundnut is

a) Epigeal
b) Hypogeal
c) Both
d) None

1969. Inside a seed, carbohydrate is stored in

a) Endosperm
b) Embryo
c) Both
d) Epidermal layer

1970. CO_2 compensation point is higher in ________ plants.

a) C_3
b) C_4
c) Both
d) None

1971. Biodiversity board was established in the year

a) 2004
b) 2008
c) 2012
d) 2016

1972. Central silk board is situated in

a) Chennai
b) Kochi
c) Banglore
d) Hyderbad

1973. Rubber board is situated in

a) Karnataka
b) Tamil Nadu
c) Kerala
d) AP

1974. Tea board is situated in

a) Kerala
b) Assam
c) West Bengal
d) Tamil Nadu

1975. IITA is situated in

a) Srilanka
b) Nigeria
c) Kenya
d) USA

1976. Saffron revolution indicates

a) Orange cultivation
b) Saffron export
c) Solar energy
d) Non-conventional energy

1977. Father of Plant Physiology

a) Haberlandt
b) Platter
c) Reiter
d) Stephen Hales

1978. Father of Agroclimatology

a) Koppen
b) Walia
c) Troll
d) De Candole

1979. Father of Agrometeorology

a) De Candole
b) Thornthwaite
c) Koppen
d) Walia

1980. Father of nitrogen fixation

a) Bauchin
b) Galton
c) Winogradsky
d) R. Mishra

1981. Growth of plant is measured by

a) Atmometer
b) Auxanometer
c) Cryometer
d) Pycnometer

1982. Leaf water potential is measured by

a) Pycnometer
b) Psycrometer
c) Drosometer
d) Pulvimeter

1983. Water stress is measured by

a) Irrometer
b) Pulvimeter
c) Cryometer
d) Hydrometer

1984. Leaf temperature is measured by

a) Pyrgeometer
b) Infrared thermometer
c) Atmometer
d) Pulvimeter

1985. Direct solar radiation is measured by

a) Pyrheliometer
b) Pyrgeometer
c) Pyradiometer
d) Spectrophotometer

1986. Moisture deficit index implies

a) (PET-P)/P
b) (P-PET)/P
c) (P-PET)/PET
d) (PET-P)/PET

1987. Moisture availability index means

a) Actual ET/Potential ET
b) Potential ET/ Actual ET
c) (P-PET)/Actual ET
d) (PET-P)/ Actual ET

1988. De Candole classified climate based on

a) Temperature
b) Rainfall
c) Moisture index
d) Vegetation

1989. Classification of climate for agriculture purpose given by

a) Koppen
b) Troll
c) Thronthwaite and Mother
d) De Candolle

1990. Koppen classified climate based on

a) Temperature
b) Rainfall
c) Evaporation
d) Both a and b

1991. Degraded land of total geographical area in India

a) 37%
b) 47%
c) 57%
d) 67%

1992. Amount of average annual surface runoff in India

a) 95 mha m
b) 105 mha m
c) 115 mha m
d) 165 mha m

1993. Compared to border irrigation, water savings under ridge and furrow method is ________ %

a) 20-30
b) 30-40
c) 25-35
d) 35-45

1994. Intercropping is recommended in areas receiving rainfall upto _____mm.

a) 800
b) 1000
c) 700
d) 900

1995. Alley cropping is recommended in the areas receiving rainfall

a) 300-450 mm
b) 400-650 mm
c) 500-750 mm
d) 600-850 mm

1996. Waterlogged area in India

a) 6.5 Mha
b) 7.5 Mha
c) 8.5 Mha
d) 9.5 Mha

1997. Dinitroaniline group of herbicide inhibits

a) Cell division
b) Respiration
c) Both a and b
d) RNA synthesis

1998. Triazine group of herbicide inhibits

a) Cell division
b) Hill reaction
c) DNA synthesis
d) All

1999. Germination is hampered by ________ group of herbicides.

a) Carbamates
b) Amides
c) Bipyridillium
d) All

2000. Cereals and pulses are deficient in _______ and _______ respectively.

a) Lycine and methionine
b) Methionine and lycine
c) Lycine and tryptophan
d) Methionine and tryptophan

2001. In arid regions, the soluble salts accumulate

a) Near the soil surface
b) Root zone
c) Both
d) None

2002. If a soil possesses the Exchangeable Sodium Percentage (ESP) greater than 15, pH more than 8.5 and Electric Conductivity (EC) below 4 ds/m, it is

a) Saline soil
b) Alkali soil
c) Acid soil
d) Saline-alkali soil

2003. Application rate of gypsum in alkali soil is

a) 1-5 t/ha
b) 5-8 t/ha
c) 8-10 t/ha
d) 10-15 t/ha

2004. Saline soil is also called

a) White alkali
b) Solonchack
c) Both a and b
d) Solanetz

2005. If a soil possesses the Exchangeable Sodium Percentage (ESP) greater than 15, pH more than 8.5 and Electric Conductivity (EC) above 4 ds/m, it is

a) Saline soil
b) Acid soil
c) Alkali soil
d) Saline-alkali soil

2006. If a soil possesses the Exchangeable Sodium Percentage (ESP) less than 15, pH less than 8.5 and Electric Conductivity (EC) above 4 ds/m, it is

a) Saline-alkali soi
b) Saline soil
c) Alkali soil
d) Acid soil

2007. Saline soil is

a) Flocculated
b) Dispersed
c) Either a or b
d) None

2008. High salt tolerant

a) Cotton b) Sesame

c) Sorghum d) Oats

2009. Low salinity tolerant crop grows in soil with EC

a) < 4 ds/m b) 4 ds/m

c) 4-10 mmhos/cm d) > 10 mmhos/cm

2010 Alkalinity tolerant crop

a) Maize b) Carrot

c) Peach d) Cotton

2011. A good cultivable soil ________ % base occupied by Ca.

a) 40 b) 50

c) 60 d) 70

2012. The greatest buffering in soil occurs at ________ % base saturation.

a) 40 b) 50

c) 60 d) 70

2013. Liming normally is recommended for soils with pH

a) <5.5 b) <7

c) >5.5 d) >8.5

2014. Boron is available at ________ pH

a) Acidic b) Alkaline

c) Neutral d) All

2015. Very low lime required for

a) Soybean b) Tobacco

c) Pineapple d) Potato

2016. Alkaline soils are deficient in

a) Zn and Mn b) Ca and Mg

c) B and Fe d) Cu and Mo

2017. Rocks are ________ when silica is present.

a) Acidic b) Alkaline

c) Neutral d) Saline

2018. If pulse crops are grown in acid soils, they will be deficient in ________

a) Mn
b) Zn
c) Mo
d) None

2019. In saline soil osmotic potential is ________

a) Less
b) Varied
c) High
d) Zero

2020. Bronzing of leaves in citrus may be due to ________ in soil.

a) Acidity
b) Alkalinity
c) Salinity
d) None

2021. Accurate temperature and humidity should be maintained in the rearing room for proper development of silkworm larvae. Name the equipment used for measuring temperature and humidity

a) Thermometer
b) Hydrometer
c) Hygrometer
d) Lactometer

2022. Which machine is preferable for reeling low quality cocoon and double cocoon?

a) Cottage basin
b) Charkha
c) Multi end reeling machine
d) Automatic reeling machine

2023. Pepper like black spots appearing the surface infected silkworm is the symptom of which diseases

a) Grasserie
b) Pebrine
c) Muscardine
d) Gattine

2024. Dupion silk obtained from

a) Good cocoon
b) Calcified cocoon
c) Double cocoon
d) Pre-mature cocoon

2025. Process of unwinding of silk filament from the cocoon with the help of reeling machine is called

a) Skeining
b) Stifling
c) Reeling
d) Riddling

2026. "CSR_2" * "CSR_4" is a hybrid variety of silkworm. The colour of cocoon is

a) Yellow
b) Pink
c) Green
d) White

2027. Silkworm are font of which light

a) Darkness
b) Dim light
c) Strong light
d) Strong light by dim light

2028. Causative organism of bacterial leaf spot disease of mulberry plant

a) Phyllactinia corylea
b) Peridospora mori
c) Pseudomonas mori
d) Fusarium solani

2029. Fungal disease of *Bombyx mori*

a) Muscardine
b) Pebrine
c) Grasserie
d) Flacherie

2030. Foundation stock seeds of silkworm eggs produced in which station?

a) P1 station
b) P2 station
c) P3 station
d) P4 station

2031. Causative organism of Flacherie disease of Bombyx mori

a) *Bacillus bombycis*
b) NPV
c) *Beaveria Bassiana*
d) *Aspergillus niger*

2032. Powdery mildew disease of mulberry plant is caused by

a) Meloidoyne Incognita
b) Phyllactina corylea
c) Cercospora moricola
d) Mealybug

2033. A local variety of mulberry plant which is low yielding but needs less fertilizer and irrigation

a) Mysore local
b) Victory-1
c) Viswa
d) :-‘“S_{36}”‘

2034. Leaf blight disease of mulberry caused by

a) Fungus
b) Nematode
c) Virus
d) Bacteria

2035. Diploid chromosome number of *Bombyx mori*

a) 24
b) 28
c) 54
d) 56

2036. Best suitable soil ‘“P^H”‘ for mulberry cultivation is ranging from

a) 8-9.5
b4) 6.2-6.8
c) 9.3-9.8
d) 4.7-5.3

2037. The Threader used in Charkha reeling machine is called

a) Croissure b) Tharppati

c) Jettebout d) Porcelain button

2038. Branch of agricultural science dealing with the principles and practices of field crop

a) Agricultural entomology b) Horticulture

c) Soil science d) Agronomy

2039. Acid treatment is used to break the diapauses of silkworm eggs. Which acid is used for acid treatment?

a) Acetic acid b) HCL

c) Sulphuric acid d) Formic acid

2040. Largest silk producers in the world

a) Brazil b) India

c) Japan d) China

2041. For rainfed mulberry cultivation plant spacing 90 cm × 90 cm, calculate number of mulberry plans required for 1 hectare area

a) 4938 b) 12345

c) 1234 d) 98767

2042. Which statement is correct for the identification of male larvae from the female?

a) Presence of herolds gland in male

b) Presence of ishiwatas gland in male

c) Presence of ''X'' marking in male

d) Large size head in male

2043. During silkworm egg production the parents of commercial layings are produced in which station

a) P1 station b) P2 station

c) P3 station d) P4 station

2044. Which state is the major producers of silk textiles in India?

a) Assam b) Tamil Nadu

c) West Bengal d) Karnataka

2045. In a grainage the equipment used at time of oviposition

a) Acid treatment batch b) Chopstick

c) Cellule d) Rotary mountage

2046. Calculate the amount of MOP (K = 60%) for 1 hector rainfed mulberry garden. If the fertilizer recommendation is 120 : 60 : 60 kg NPK/ha

a) 100 kg b) 200 kg

c) 300 kg d) 400 kg

2047. Number of cocoons in 1 kg is 560Pierced cocoon-5, double cocoons-15, flimsy cocoons-8, malformed cocoons-4 Calculate the cocoon percentage

a) 85% b) 95.7%

c) 90.4% d) 94.3%

2048. Pollination in mulberry is

a) Anemophily b) Zoophily

c) Hydrophily d) Entomophily

2049. If fertilizer recommendation is 300 : 120 : 120, calculate the amount of urea required for 1 hector mulberry garden

a) 1500 kg b) 652 kg

c) 1500 kg d) 266 kg

2050. ______ is a commonly used root hormone for propagating mulberry plant.

a) Gibberellins b) IAA

c) Ethylene d) Abscisic acid

2051. Queen of textiles

a) Cotton b) Cocoon

c) Silk d) Kanchipuram Sarees

2052. Name the proteins by which silk is made of

a) Fibroin b) Sericine

c) Albumin d) Globulin

2053. In tropical region the best orientation of long axis of silkworm rearing house is

a) East west b) North South

c) North west and South east d) North east and South west

2054. Which rearing equipment is not used at the time of spinning of silkworm larvae?

a) Chandrika b) Feeding stand
c) Netrika d) Rotary mountage

2055. Viral disease of silkworm is

a) Tukra b) Muscuardine
c) Grasserie d) Pebrine

2056. Which is not a part of silkworm egg?

a) Filippi's gland b) Chorion
c) Yolk d) Veitelline membrane

2057. The process of shedding down of exoskeleton in silkworm is called

a) Mulching b) Moulting
c) Brushing d) Mounting

2058. Ahimsa silk is naturally obtained from

a) Mulberry silk b) Tasar silk
c) Muga silk d) Eri silk

2059. Which district the silkworm eggs are produced in Kerala?

a) Kasargod b) Kannur
c) Wayanad d) Palakkad

2060 Mother moth examination is conducted for the detection of which disease

a) Pebrine b) Flacherie
c) Grasserie d) Muscardine

2061. Example for a phosphate solubilizing microbial bio-fertilizer

a) Ankush b) SSP
c) Seriphose d) Seriboost

2062. Ram Nagar is the famous cocoon market in Asia. Where is it located?

a) Karnataka b) Tamil Nadu
c) West Bengal d) Assam

2063. During silk test which test is done to evaluate the evenness, cleanness and neatness defects of raw silk?

a) Tenacity and elongation test b) Size test
c) Seriplane test d) Cohesion test

2064. 'Voltinism' is referred on the basis of

a) Number of larval moulting b) Geographical distribution

c) Number of generation per year d) None of these

2065. 'Indian Silk' is the complete sericulture and silk industry journal of India, published since 1962. It is published by

a) SERIFED b) CSR and TI

c) NSP d) CSB

2066. Serious pest of mulberry plant is

a) Bihar hairy caterpillar b) Uzifly

c) Tukra d) Dermistid beetle

2067. Cocoon weight of '"CSR_2"' races of silkworm is ranging from

a) 3.3-3.5 g b) 1.3-1.5 g

c) 1.8-1.9 g d) 2.5-2.9 g

2068. *Bombyx mori* complete their life cycle within

a) 30-35 days b) 60-65 days

c) 90 days d) 3 weeks

2069. "Sampoorna' is used for

a) To trap and kill uziflies

b) Bed disinfectant to prevent silkworm diseases

c) Uniform maturation of silkworms

d) Micro nutritional foliary spray for mulberry

2070. Bacterial disease of mulberry plant

a) Leaf spot b) Leaf rust

c) Powdery mildew d) Root rot

2071. Mulberry belongs to the family

a) *Malvaceae* b) *Moraceae*

c) *Apcoynaceae* d) *Rubiaceae*

2072. Food plant of Muga Silkworm is

a) Som and Soalu b) Mulberry

c) Castor d) Asan and Arjun

2073. Central Silk Board is the central authority to deal with sericulture activities in India. Where is its headquarters located?

a) Bangalore
b) New Delhi
c) Mysore
d) Chennai

2074. Example for room disinfectant used in rearing house

a) Ankush
b) Vijetha power
c) RKO power
d) Bleaching powder

2075. Raw silk percentage of CSR races ranges from

a) 18-20
b) 23-25
c) 26-29
d) >60

2076. Use of silk was discovered by

a) China
b) India
c) Brazil
d) Sri Lanka

2077. The number of kilograms of cocoons required to obtain 1 kg reeled silk is called

a) Denier
b) Renditta
c) Kakame cost
d) Shell ratio

2078. Scientific name of mulberry silk worm

a) *Antheraea assamensis*
b) *Philosamia ricini*
c) *Bombyx mori*
d) *Antheraea mylitta*

2079. Study of mulberry cultivation

a) Pisciculture
b) Sericulture
c) Moriculture
d) Apiculture

2080. The protozoan diseases of silkworm

a) Pebrine
b) Muscardine
c) Flacherie
d) Grasserie

2081. Fertilizer recommendation for irrigated mulberry cultivation

a) 140:55:275
b) 225:75:75
c) 400:200:400
d) 300:120:120

2082. Given below is a sequence of steps in the processing of reeling. Which are the missing steps. Stifling __________ reeling ____________

a) Cooking, re-reeling
b) Lacing, skeining
c) Lacing, cooking
d) Sorting, brushing

2083. In India the purity of silk ensures through the trade mark

a) ISI mark b) Agmark

c) Silk mark d) Seri mark

2084. The unit used to measure the thickness of silk filament

a) Renditta b) Denier

c) Micron d) MM

2085. Industrial Seed of Silkworm (DFLs) are produced in

a) Hatcheries b) Grainage

c) Breeding station d) Breeding centers

2086. Silk obtained from

a) Pupa b) Larva

c) Cocoon d) Moth

2087. Diploid chromosome number of mulberry plant

a) 28 b) 24

c) 54 d) 56

2088. In South India mulberry propagation is mainly through which method?

a) Grafting b) Layering

c) Cuttings d) Budding

2089. The number of kilograms of cocoons required to obtain 1 kg reeled silk is called

a) Shell ratio b) Kakame cost

c) Renditta d) Denier

2090. Which is not a silkworm seed?

a) ‘‘“BL_{23}”’‘ b) ‘‘“S_{34}”’‘

c) ‘‘“NB_4 D_2”’‘ d) ‘‘“BL_{24}”’‘

2091. Inflorescence elongation is due to the presence of which enzyme?

a) Auxin b) Gibberellin

c) Cytokinin d) None of these

2092. Which portion of leaves contribute much to ear growth?

a) Top b) Middle

c) Bottom d) None of these

2093. Fertilizer schedule includes

a) Dose only

b) Source only

a) Time of application

d) Dose, source, time and method of application

2094. N-P-K proportion in balanced fertilization for most of the crops is:

a) 4 : 1 : 1
b) 2 : 1 : 1
c) 6 : 2 : 1
d) 6 : 3 : 1

2095. Normally FYM is applied to the field:

a) One day before sowing
b) 7 days before sowing
c) 15 days before sowing
d) 30 days before sowing

2096. N-P-K proportion in balanced fertilization for pulse crops is:

a) 4 : 1 : 1
b) 1 : 2 : 2
c) 6 : 2 : 1
d) 6 : 3 : 1

2097. In conservation Agriculture minimum amount of crop residue remains on soil:

a) 10%
b) 15%
c) 25%
d) 30%

2098. Which type of plants can utilize the CO_2 even at low concentration?

a) C_3
b) C_4
c) CAM
d) All of these

2099. Blue green algae fix nitrogen mainly in :

a) Rice field
b) Cereals, field
c) Sugarcane field
d) Legumes, field

2100. The nutrient responsible for development of root is:

a) N
b) S
c) Zn
d) P

Answers

Agro-meteorology, Crop Management, Soil Management, Nutrient Management, Water Management, Weed Management, Dryland Agriculture and Agroforestry

1	d	31	a	61	a	91	b	121	c
2	c	32	b	62	a	92	b	122	a
3	c	33	d	63	d	93	a	123	c
4	c	34	c	64	a	94	a	124	a
5	b	35	b	65	a	95	c	125	a
6	b	36	a	66	d	96	a	126	a
7	c	37	b	67	a	97	a	127	a
8	c	38	b	68	a	98	a	128	b
9	d	39	d	69	b	99	c	129	b
10	c	40	b	70	a	100	b	130	a
11	b	41	c	71	d	101	a	131	b
12	b	42	d	72	a	102	d	132	c
13	d	43	a	73	a	103	a	133	c
14	d	44	d	74	d	104	a	134	c
15	b	45	a	75	c	105	b	135	d
16	d	46	c	76	a	106	d	136	a
17	c	47	c	77	b	107	b	137	a
18	d	48	b	78	a	108	c	138	c
19	b	49	a	79	d	109	a	139	a
20	c	50	c	80	a	110	a	140	b
21	d	51	c	81	a	111	a	141	c
22	a	52	a	82	c	112	b	142	a
23	b	53	b	83	b	113	c	143	b
24	d	54	a	84	c	114	a	144	a
25	b	55	a	85	c	115	b	145	a
26	b	56	c	86	b	116	b	146	c
27	a	57	b	87	b	117	b	147	b
28	d	58	d	88	b	118	a	148	a
29	c	59	c	89	b	119	d	149	a
30	d	60	b	90	a	120	c	150	a

151	a	189	b	227	d	265	b	303	b
152	a	190	b	228	c	266	a	304	b
153	a	191	b	229	a	267	b	305	b
154	a	192	a	230	a	268	c	306	a
155	a	193	d	231	b	269	d	307	c
156	b	194	a	232	c	270	a	308	d
157	c	195	d	233	a	271	b	309	b
158	d	196	b	234	a	272	d	310	a
159	b	197	a	235	a	273	c	311	d
160	b	198	b	236	c	274	a	312	b
161	d	199	b	237	a	275	a	313	b
162	a	200	b	238	a	276	a	314	b
163	a	201	a	239	c	277	a	315	c
164	a	202	b	240	b	278	a	316	c
165	c	203	d	241	c	279	a	317	a
166	a	204	a	242	b	280	d	318	b
167	b	205	b	243	b	281	b	319	a
168	a	206	a	244	c	282	a	320	b
169	b	207	a	245	c	283	a	321	a
170	a	208	c	246	c	284	a	322	a
171	b	209	a	247	c	285	b	323	c
172	a	210	a	248	d	286	a	324	d
173	b	211	a	249	b	287	b	325	b
174	b	212	a	250	a	288	b	326	b
175	c	213	c	251	b	289	d	327	c
176	c	214	a	252	c	290	b	328	a
177	b	215	b	253	b	291	d	329	c
178	b	216	a	254	a	292	b	330	d
179	a	217	a	255	c	293	a	331	a
180	a	218	c	256	b	294	a	332	a
181	d	219	c	257	c	295	a	333	b
182	b	220	a	258	b	296	b	334	b
183	a	221	a	259	a	297	c	335	a
184	b	222	a	260	c	298	d	336	c
185	c	223	a	261	a	299	b	337	c
186	a	224	a	262	a	300	d	338	a
187	a	225	c	263	b	301	b	339	d
188	c	226	a	264	a	302	b	340	a

341	b	379	c	417	a	455	d	493	a
342	b	380	d	418	b	456	a	494	b
343	b	381	d	419	c	457	a	495	c
344	d	382	c	420	b	458	a	496	a
345	c	383	c	421	a	459	b	497	b
346	d	384	d	422	a	460	d	498	b
347	c	385	a	423	d	461	d	499	c
348	a	386	c	424	b	462	c	500	a
349	c	387	d	425	a	463	b	501	a
350	b	388	b	426	a	464	a	502	b
351	a	389	a	427	d	465	b	503	c
352	b	390	c	428	a	466	b	504	d
353	c	391	c	429	a	467	c	505	c
354	d	392	a	430	a	468	a	506	c
355	a	393	d	431	a	469	b	507	a
356	d	394	b	432	a	470	a	508	a
357	a	395	a	433	d	471	b	509	a
358	b	396	b	434	a	472	a	510	c
359	b	397	a	435	b	473	b	511	a
360	c	398	b	436	a	474	a	512	b
361	c	399	a	437	b	475	a	513	c
362	c	400	b	438	b	476	c	514	d
363	a	401	b	439	d	477	d	515	a
364	d	402	d	440	a	478	a	516	d
365	c	403	b	441	a	479	c	517	b
366	a	404	c	442	b	480	d	518	c
367	b	405	b	443	c	481	b	519	a
368	c	406	b	444	d	482	c	520	c
369	a	407	a	445	d	483	b	521	a
370	a	408	a	446	b	484	b	522	a
371	b	409	b	447	d	485	c	523	c
372	a	410	d	448	d	486	a	524	b
373	a	411	d	449	d	487	b	525	b
374	a	412	a	450	a	488	c	526	d
375	d	413	c	451	d	489	c	527	a
376	a	414	c	452	d	490	a	528	b
377	b	415	c	453	c	491	d	529	b
378	b	416	a	454	d	492	a	530	d

531	a	569	a	607	a	645	c	683	b
532	b	570	d	608	b	646	a	684	c
533	c	571	b	609	a	647	d	685	c
534	a	572	d	610	d	648	a	686	b
535	a	573	b	611	a	649	b	687	d
536	b	574	a	612	c	650	b	688	b
537	c	575	c	613	b	651	a	689	c
538	b	576	a	614	c	652	a	690	c
539	a	577	c	615	b	653	a	691	d
540	b	578	a	616	b	654	a	692	c
541	c	579	c	617	c	655	c	693	d
542	b	580	b	618	c	656	d	694	a
543	a	581	a	619	a	657	c	695	a
544	b	582	b	620	c	658	c	696	a
545	b	583	a	621	d	659	c	697	b
546	d	584	b	622	a	660	b	698	b
547	a	585	a	623	b	661	b	699	a
548	b	586	b	624	a	662	c	700	b
549	c	587	d	625	a	663	a	701	a
550	b	588	a	626	b	664	d	702	b
551	d	589	d	627	d	665	b	703	b
552	d	590	a	628	d	666	a	704	c
553	b	591	b	629	c	667	c	705	c
554	d	592	c	630	b	668	a	706	b
555	b	593	a	631	d	669	a	707	a
556	c	594	b	632	c	670	b	708	a
557	b	595	a	633	a	671	c	709	d
558	a	596	b	634	b	672	a	710	b
559	b	597	b	635	c	673	b	711	a
560	d	598	a	636	c	674	b	712	c
561	a	599	c	637	c	675	c	713	a
562	d	600	c	638	b	676	c	714	c
563	a	601	b	639	a	677	b	715	a
564	a	602	a	640	a	678	b	716	b
565	b	603	d	641	c	679	c	717	c
566	b	604	d	642	b	680	b	718	a
567	c	605	c	643	b	681	d	719	b
568	c	606	d	644	d	682	d	720	b

721	a	759	a	797	c	835	b	873	a
722	c	760	b	798	b	836	c	874	b
723	b	761	a	799	c	837	a	875	d
724	b	762	c	800	a	838	b	876	d
725	a	763	b	801	b	839	a	877	a
726	b	764	a	802	a	840	b	878	b
727	b	765	a	803	c	841	c	879	b
728	b	766	b	804	b	842	a	880	c
729	a	767	c	805	b	843	a	881	a
730	a	768	d	806	b	844	a	882	c
731	b	769	a	807	a	845	c	883	c
732	a	770	b	808	d	846	d	884	a
733	b	771	a	809	d	847	a	885	b
734	b	772	b	810	b	848	d	886	a
735	b	773	a	811	a	849	a	887	b
736	d	774	d	812	b	850	d	888	a
737	b	775	d	813	a	851	a	889	a
738	a	776	a	814	c	852	d	890	b
739	c	777	a	815	d	853	b	891	b
740	a	778	a	816	b	854	d	892	b
741	d	779	a	817	a	855	a	893	b
742	b	780	d	818	a	856	b	894	a
743	c	781	c	819	c	857	d	895	c
744	a	782	a	820	a	858	c	896	b
745	b	783	d	821	b	859	a	897	a
746	a	784	b	822	b	860	d	898	a
747	a	785	c	823	b	861	b	899	c
748	a	786	b	824	a	862	a	900	b
749	a	787	a	825	b	863	b	901	a
750	b	788	a	826	a	864	a	902	c
751	c	789	c	827	b	865	b	903	b
752	a	790	b	828	c	866	a	904	a
753	a	791	b	829	b	867	b	905	a
754	c	792	a	830	a	868	a	906	c
755	b	793	d	831	b	869	d	907	b
756	a	794	b	832	a	870	a	908	a
757	a	795	a	833	b	871	b	909	b
758	b	796	b	834	a	872	a	910	a

911	a	949	a	987	b	1025	a	1063	c
912	b	950	c	988	a	1026	a	1064	c
913	a	951	d	989	a	1027	d	1065	d
914	d	952	a	990	a	1028	a	1066	a
915	d	953	b	991	b	1029	a	1067	b
916	b	954	d	992	a	1030	a	1068	c
917	d	955	a	993	a	1031	a	1069	b
918	a	956	c	994	b	1032	a	1070	c
919	b	957	c	995	b	1033	b	1071	a
920	c	958	b	996	a	1034	c	1072	c
921	c	959	a	997	a	1035	c	1073	c
922	b	960	b	998	a	1036	d	1074	d
923	c	961	d	999	b	1037	c	1075	a
924	d	962	d	1000	b	1038	b	1076	d
925	d	963	b	1001	b	1039	c	1077	c
926	a	964	a	1002	c	1040	b	1078	d
927	b	965	b	1003	b	1041	c	1079	c
928	a	966	b	1004	a	1042	a	1080	a
929	a	967	a	1005	c	1043	d	1081	a
930	d	968	a	1006	d	1044	c	1082	d
931	a	969	b	1007	c	1045	b	1083	b
932	a	970	a	1008	a	1046	c	1084	c
933	a	971	b	1009	a	1047	d	1085	a
934	c	972	b	1010	b	1048	c	1086	c
935	d	973	d	1011	b	1049	c	1087	a
936	a	974	a	1012	b	1050	d	1088	b
937	a	975	b	1013	c	1051	d	1089	a
938	b	976	b	1014	b	1052	b	1090	a
939	d	977	a	1015	a	1053	b	1091	b
940	a	978	c	1016	b	1054	b	1092	d
941	c	979	c	1017	b	1055	b	1093	a
942	a	980	b	1018	a	1056	a	1094	a
943	a	981	b	1019	a	1057	b	1095	a
944	b	982	b	1020	b	1058	a	1096	a
945	c	983	b	1021	b	1059	a	1097	b
946	a	984	c	1022	a	1060	b	1098	d
947	a	985	a	1023	c	1061	a	1099	c
948	a	986	b	1024	b	1062	b	1100	b

1101	c	1139	d	1177	b	1215	b	1253	d
1102	c	1140	d	1178	c	1216	c	1254	c
1103	a	1141	d	1179	c	1217	c	1255	c
1104	a	1142	c	1180	c	1218	a	1256	c
1105	a	1143	a	1181	b	1219	b	1257	d
1106	b	1144	d	1182	c	1220	b	1258	c
1107	b	1145	d	1183	d	1221	a	1259	c
1108	a	1146	b	1184	b	1222	b	1260	c
1109	a	1147	d	1185	d	1223	a	1261	d
1110	b	1148	c	1186	b	1224	a	1262	d
1111	a	1149	d	1187	c	1225	b	1263	c
1112	c	1150	a	1188	c	1226	b	1264	d
1113	c	1151	a	1189	b	1227	a	1265	b
1114	c	1152	b	1190	b	1228	a	1266	b
1115	d	1153	d	1191	b	1229	d	1267	b
1116	c	1154	c	1192	a	1230	c	1268	c
1117	c	1155	a	1193	b	1231	d	1269	c
1118	d	1156	d	1194	b	1232	d	1270	c
1119	b	1157	c	1195	c	1233	d	1271	c
1120	a	1158	a	1196	a	1234	c	1272	c
1121	d	1159	d	1197	b	1235	c	1273	d
1122	d	1160	c	1198	b	1236	d	1274	d
1123	c	1161	c	1199	c	1237	c	1275	d
1124	b	1162	a	1200	b	1238	d	1276	c
1125	d	1163	b	1201	a	1239	d	1277	d
1126	a	1164	b	1202	c	1240	d	1278	d
1127	c	1165	a	1203	b	1241	d	1279	c
1128	d	1166	b	1204	c	1242	d	1280	b
1129	d	1167	d	1205	d	1243	d	1281	d
1130	b	1168	a	1206	c	1244	d	1282	b
1131	c	1169	c	1207	c	1245	c	1283	a
1132	a	1170	a	1208	b	1246	d	1284	b
1133	b	1171	a	1209	d	1247	a	1285	a
1134	d	1172	c	1210	d	1248	d	1286	b
1135	c	1173	b	1211	b	1249	d	1287	a
1136	d	1174	a	1212	b	1250	a	1288	c
1137	d	1175	b	1213	b	1251	b	1289	d
1138	b	1176	c	1214	b	1252	c	1290	c

1291	d	1329	c	1367	b	1405	a	1443	c
1292	c	1330	a	1368	c	1406	a	1444	b
1293	a	1331	a	1369	a	1407	a	1445	b
1294	d	1332	a	1370	c	1408	a	1446	d
1295	a	1333	c	1371	d	1409	b	1447	c
1296	d	1334	a	1372	c	1410	a	1448	a
1297	a	1335	c	1373	d	1411	a	1449	a
1298	c	1336	c	1374	a	1412	a	1450	c
1299	a	1337	d	1375	a	1413	c	1451	d
1300	b	1338	a	1376	d	1414	a	1452	c
1301	a	1339	d	1377	a	1415	a	1453	d
1302	b	1340	c	1378	b	1416	a	1454	b
1303	d	1341	b	1379	a	1417	a	1455	c
1304	a	1342	a	1380	d	1418	b	1456	c
1305	d	1343	d	1381	a	1419	b	1457	b
1306	d	1344	d	1382	d	1420	b	1458	a
1307	d	1345	b	1383	a	1421	a	1459	d
1308	c	1346	a	1384	c	1422	a	1460	b
1309	c	1347	a	1385	d	1423	a	1461	d
1310	c	1348	c	1386	c	1424	a	1462	c
1311	b	1349	b	1387	d	1425	a	1463	d
1312	b	1350	a	1388	c	1426	b	1464	d
1313	d	1351	b	1389	a	1427	b	1465	c
1314	a	1352	b	1390	c	1428	a	1466	d
1315	d	1353	c	1391	d	1429	c	1467	d
1316	c	1354	d	1392	a	1430	b	1468	d
1317	c	1355	a	1393	a	1431	c	1469	d
1318	a	1356	a	1394	d	1432	b	1470	c
1319	a	1357	b	1395	b	1433	a	1471	d
1320	a	1358	b	1396	d	1434	a	1472	d
1321	d	1359	c	1397	a	1435	a	1473	d
1322	c	1360	a	1398	b	1436	c	1474	d
1323	a	1361	d	1399	d	1437	d	1475	d
1324	a	1362	b	1400	a	1438	b	1476	d
1325	a	1363	a	1401	a	1439	c	1477	d
1326	a	1364	d	1402	a	1440	d	1478	d
1327	d	1365	d	1403	c	1441	b	1479	d
1328	a	1366	b	1404	a	1442	a	1480	b

1481	b	1519	c	1557	d	1595	d	1633	c
1482	d	1520	c	1558	b	1596	c	1634	b
1483	d	1521	c	1559	b	1597	d	1635	d
1484	c	1522	b	1560	c	1598	a	1636	a
1485	d	1523	d	1561	b	1599	c	1637	a
1486	a	1524	c	1562	b	1600	b	1638	a
1487	b	1525	c	1563	a	1601	c	1639	b
1488	d	1526	b	1564	b	1602	b	1640	a
1489	c	1527	d	1565	b	1603	a	1641	b
1490	d	1528	d	1566	d	1604	c	1642	a
1491	c	1529	b	1567	b	1605	a	1643	b
1492	d	1530	c	1568	d	1606	a	1644	a
1493	d	1531	b	1569	b	1607	a	1645	a
1494	a	1532	d	1570	b	1608	d	1646	b
1495	c	1533	d	1571	b	1609	b	1647	c
1496	c	1534	c	1572	d	1610	b	1648	d
1497	b	1535	d	1573	b	1611	a	1649	b
1498	d	1536	d	1574	d	1612	b	1650	c
1499	c	1537	b	1575	d	1613	b	1651	b
1500	d	1538	a	1576	b	1614	a	1652	b
1501	c	1539	d	1577	b	1615	b	1653	c
1502	a	1540	c	1578	d	1616	c	1654	b
1503	b	1541	c	1579	c	1617	b	1655	c
1504	a	1542	b	1580	d	1618	a	1656	b
1505	d	1543	a	1581	d	1619	c	1657	d
1506	a	1544	c	1582	b	1620	c	1658	a
1507	b	1545	d	1583	b	1621	a	1659	a
1508	a	1546	c	1584	c	1622	a	1660	b
1509	a	1547	c	1585	a	1623	c	1661	a
1510	a	1548	b	1586	d	1624	c	1662	c
1511	c	1549	b	1587	b	1625	d	1663	a
1512	c	1550	d	1588	a	1626	b	1664	b
1513	d	1551	d	1589	a	1627	c	1665	a
1514	b	1552	d	1590	d	1628	d	1666	B
1515	c	1553	c	1591	b	1629	b	1667	a
1516	b	1554	c	1592	c	1630	b	1668	c
1517	c	1555	a	1593	b	1631	d	1669	d
1518	c	1556	d	1594	d	1632	c	1670	a

1671	d	1709	b	1747	d	1785	a	1823	a
1672	c	1710	d	1748	d	1786	c	1824	c
1673	b	1711	a	1749	d	1787	d	1825	c
1674	d	1712	d	1750	b	1788	c	1826	b
1675	c	1713	b	1751	c	1789	d	1827	c
1676	b	1714	c	1752	b	1790	b	1828	a
1677	c	1715	d	1753	b	1791	b	1829	b
1678	b	1716	d	1754	d	1792	c	1830	b
1679	c	1717	b	1755	d	1793	b	1831	b
1680	a	1718	c	1756	d	1794	c	1832	c
1681	c	1719	b	1757	d	1795	d	1833	a
1682	d	1720	c	1758	a	1796	d	1834	c
1683	b	1721	c	1759	d	1797	a	1835	b
1684	d	1722	d	1760	c	1798	d	1836	c
1685	d	1723	c	1761	a	1799	a	1837	b
1686	c	1724	d	1762	a	1800	b	1838	c
1687	a	1725	b	1763	a	1801	c	1839	b
1688	d	1726	d	1764	c	1802	c	1840	a
1689	d	1727	d	1765	c	1803	b	1841	a
1690	b	1728	d	1766	c	1804	a	1842	d
1691	b	1729	b	1767	b	1805	a	1843	b
1692	b	1730	b	1768	d	1806	b	1844	c
1693	a	1731	b	1769	a	1807	a	1845	b
1694	d	1732	c	1770	b	1808	b	1846	b
1695	c	1733	d	1771	c	1809	b	1847	d
1696	d	1734	b	1772	a	1810	d	1848	d
1697	b	1735	b	1773	c	1811	a	1849	b
1698	b	1736	b	1774	a	1812	b	1850	c
1699	c	1737	d	1775	a	1813	a	1851	b
1700	a	1738	d	1776	c	1814	b	1852	a
1701	b	1739	a	1777	d	1815	a	1853	c
1702	c	1740	b	1778	c	1816	c	1854	b
1703	c	1741	d	1779	d	1817	c	1855	c
1704	b	1742	d	1780	a	1818	b	1856	d
1705	d	1743	c	1781	c	1819	c	1857	b
1706	a	1744	a	1782	a	1820	d	1858	b
1707	c	1745	c	1783	b	1821	b	1859	a
1708	d	1746	b	1784	c	1822	a	1860	b

1861	a	1899	b	1937	b	1975	b	2013	a
1862	c	1900	a	1938	c	1976	c	2014	c
1863	b	1901	d	1939	c	1977	d	2015	c
1864	d	1902	d	1940	b	1978	a	2016	a
1865	c	1903	a	1941	b	1979	d	2017	a
1866	b	1904	c	1942	d	1980	c	2018	c
1867	b	1905	b	1943	d	1981	a	2019	a
1868	c	1906	d	1944	a	1982	b	2020	b
1869	b	1907	c	1945	c	1983	a	2021	c
1870	b	1908	b	1946	b	1984	a	2022	b
1871	a	1909	a	1947	c	1985	a	2023	b
1872	a	1910	d	1948	a	1986	c	2024	c
1873	b	1911	a	1949	d	1987	a	2025	c
1874	b	1912	d	1950	b	1988	d	2026	d
1875	c	1913	a	1951	b	1989	b	2027	b
1876	b	1914	b	1952	b	1990	d	2028	c
1877	b	1915	c	1953	d	1991	c	2029	a
1878	a	1916	d	1954	b	1992	c	2030	b
1879	c	1917	b	1955	b	1993	b	2031	a
1880	c	1918	a	1956	d	1994	a	2032	b
1881	c	1919	b	1957	b	1995	c	2033	a
1882	b	1920	b	1958	a	1996	c	2034	d
1883	a	1921	b	1959	c	1997	c	2035	d
1884	d	1922	b	1960	a	1998	b	2036	b
1885	c	1923	d	1961	b	1999	b	2037	b
1886	d	1924	a	1962	b	2000	a	2038	d
1887	b	1925	a	1963	c	2001	c	2039	b
1888	c	1926	c	1964	c	2002	b	2040	d
1889	b	1927	c	1965	a	2003	b	2041	b
1890	b	1928	b	1966	d	2004	c	2042	a
1891	b	1929	b	1967	c	2005	d	2043	a
1892	b	1930	d	1968	a	2006	b	2044	d
1893	b	1931	a	1969	a	2007	a	2045	c
1894	c	1932	c	1970	a	2008	d	2046	a
1895	c	1933	d	1971	b	2009	c	2047	d
1896	b	1934	a	1972	c	2010	d	2048	a
1897	a	1935	a	1973	c	2011	c	2049	b
1898	d	1936	d	1974	c	2012	b	2050	d

2051	c	2061	c	2071	b	2081	d	2091	b
2052	a	2062	a	2072	a	2082	a	2092	a
2053	b	2063	c	2073	a	2083	c	2093	d
2054	b	2064	c	2074	d	2084	b	2094	b
2055	c	2065	d	2075	d	2085	b	2095	d
2056	a	2066	a	2076	b	2086	c	2096	b
2057	b	2067	c	2077	b	2087	a	2097	d
2058	d	2068	b	2078	c	2088	c	2098	b
2059	d	2069	c	2079	c	2089	c	2099	a
2060	a	2070	a	2080	a	2090	b	2100	d